Arthur J. Lembo, Jr.

2014

Copyright © 2014 by Arthur J. Lembo, Jr.

All rights reserved. This book or any portion thereof may not be reproduced or used in any manner whatsoever without the express written permission of the author except for the use of brief quotations in a book review or scholarly journal.

First Printing: 2014

Arthur J. Lembo, Jr.
1440 East Sandy Acres Drive
Salisbury, MD 21804

www.artlembo.wordpress.com

Table of Contents

(following the chapter titles from An Introduction to Statistical Problem Solving in Geography by Waveland Press)

Forward 3
Chapter 1 Introduction: The Context of Statistical Techniques 1
Chapter 2 Geographic Data: Characteristics and Preparation 9
Chapter 3 Descriptive Statistics and Graphics 17
Chapter 4 Descriptive Spatial Statistics 21
Chapter 5 Basic Probability and Discreet Probability Distributions 25
Chapter 6 Continuous Probability Distributions and Probability Mapping 33
Chapter 7 Basic Elements of Sampling 39
Chapter 8 Estimation in Sampling 47
Chapter 9 Elements of Inferential Statistics 53
Chapter 10 Two Sample and Dependent-Sample (Matched-Pairs) Difference Tests 61
Chapter 11 Three or More Sample Difference Tests (ANOVA) 69
Chapter 12 Categorical Difference Tests 75
Chapter 14 Point Pattern Analysis 79
Chapter 15 Area Pattern Analysis 89
Chapter 16 Correlation Analysis 93
Chapters 17 and 18 Regression Analysis 97

Forward

The purpose of this workbook is to provide a set of exercises that reinforce the concepts of statistical problem solving in geography. This workbook is used in my course *GEOG 204 Statistical Problem Solving in Geography*, and follows the chapters in the textbook *An Introduction to Statistical Problem Solving in Geography* published by Waveland Press. Similar to the textbook, my goal was to create a set of exercises that are at their heart, *geographically based*. I have found that geography students have an easier time understanding statistical concepts when the examples they explore are geographic in nature. Further, the examples provide the undergraduate geographer a glimpse into many different sub-topics within geography such as environmental and atmospheric science, epidemiology, crime, agronomy, natural resource assessment and urban planning.

All of the exercises were field tested in the aforementioned course to assess whether it was appropriate for the undergraduate geographer. I have to give a special thanks to Dr. Daniel Harris of Salisbury University who enthusiastically introduced each chapter to his students, sometimes only hours after I completed them. Nonetheless, even with the help of others, any mistakes within this exercise book are my own, and I am eager to fix any mistakes brought to my attention.

A note to professors: Field testing of the workbook proved that there are simply too many exercises to complete in a single semester. Therefore, when using this workbook, the instructor has the flexibility to pick and choose specific questions that the students should answer. In some instances you may wish to use certain questions in your lecture examples. Also, the data and workbook were designed to be completed by hand, although even in our own courses we provide opportunities for the students to use statistical software such as Excel, Minitab, or R. An answer key is available to any instructor using this book in their course; simply contact me at the email below with proof of your role as the Professor of record for the course.

Finally, this book is supplied as a *print-on-demand* book so that I can continually make updates to the material – in that vein, I consider this to be a living document with fresh examples from the many fields in geography. Finally, the use of a print-on-demand provider allows the book to be created at a minimal cost. Should you want to learn more about the concepts of statistical problem solving in geography, GIS, and spatial concepts, please visit my blog at www.artlembo.wordpress.com. Data for all the examples can be downloaded from:

http://faculty.salisbury.edu/~ajlembo/geogworkbook.xls

Arthur J. Lembo, Jr., November 2014
artlembo@gmail.com

Chapter 1
Introduction:
The Context of Statistical Techniques

Major Goals and Objectives of this Chapter:

In Chapter 1 of *An Introduction to Statistical Problem Solving in Geography* you learned to identify the types of questions geographers ask and were introduced to the importance and uses of statistics in the geographic research process, especially as it relates to hypothesis formulation. For these exercises you will formulate possible descriptive statements while evaluating spatial patterns.

Problems and Exercises:

As discussed in the textbook (Sections 1.1 and 1.2), the process of geographic problem solving and scientific inquiry often begins with the identification of an interesting and relevant spatial pattern. Oftentimes, these patterns are provided in map form, such as the five choropleth maps of the countries of the world, shown at the end of the chapter.

The data for the maps were obtained from the CIA World Factbook[1], and show (1) Live births per 1000 in population; (2) Gross Domestic Product Per Capita; (3) Percent of Total Population Classified as Urban; (4) The Annual Electricity Generated, Expressed in Kilowatt-Hours; and (5) Life Expectancy at Birth. The data for these maps are found in the accompanying spreadsheet under *Chapter 1 – World Statistics*.

For the following questions, refer to Section 1.1 in your textbook and review how the authors identified relevant questions related to the spatial patterns presented in the book. Use the discussion in the textbook to identify pertinent geographically relevant questions and observations.

---

[1] The World Factbook is accessible at https://www.cia.gov/library/publications/the-world-factbook/index.html, and was last accessed on August 8, 2014.

Select one of the five mapped variables. List three "where", "why", and "what-to-do" questions that geographers could pose to examine or investigate this spatial pattern.

Map Variable: _____

1. Briefly describe the spatial pattern (where) of the variable you selected, making sure you mention important world regions and countries in your descriptions.

2. Propose some reasons **why** you think the observed spatial pattern varies in the way it does. List some of the locational processes that you believe operate to create the pattern. What factors can you suggest to help explain the nature of the spatial distribution?

3. Create a list of at least three possible hypotheses related to the variable you selected. Be as clear and specific as possible. As a guide, you might want to reexamine the lists of hypotheses associated with the examples in chapter 1 (obesity levels for the United States, Life Expectancy at Birth (2009), Last Spring Frost dates, and Population Change in the United States).

a.

b.

c.

Reexamine the four maps which you did not select for analysis in the previous steps and chose **one**. Do you see a spatial relationship between the map you chose first and your newly-selected map? That is, do the spatial patterns of the two maps seem similar? *For example, does there appear to be a relationship between countries with high electricity use and per capita GDP? Or, does low electricity use seem to have an influence on the life expectancy of the population?* Describe the nature and extent of this spatial relationship, referring to particular world regions (i.e. *Africa, Europe*) or world categories (i.e. *less developed*) as appropriate. What explanation can you provide for this relationship?

Variable 2: _____

4. Using the two mapped variables you selected, suggest at least three hypotheses geographers could form to explore potential relationships.

a.

b.

c.

Workbook for Statistical Problem Solving in Geography

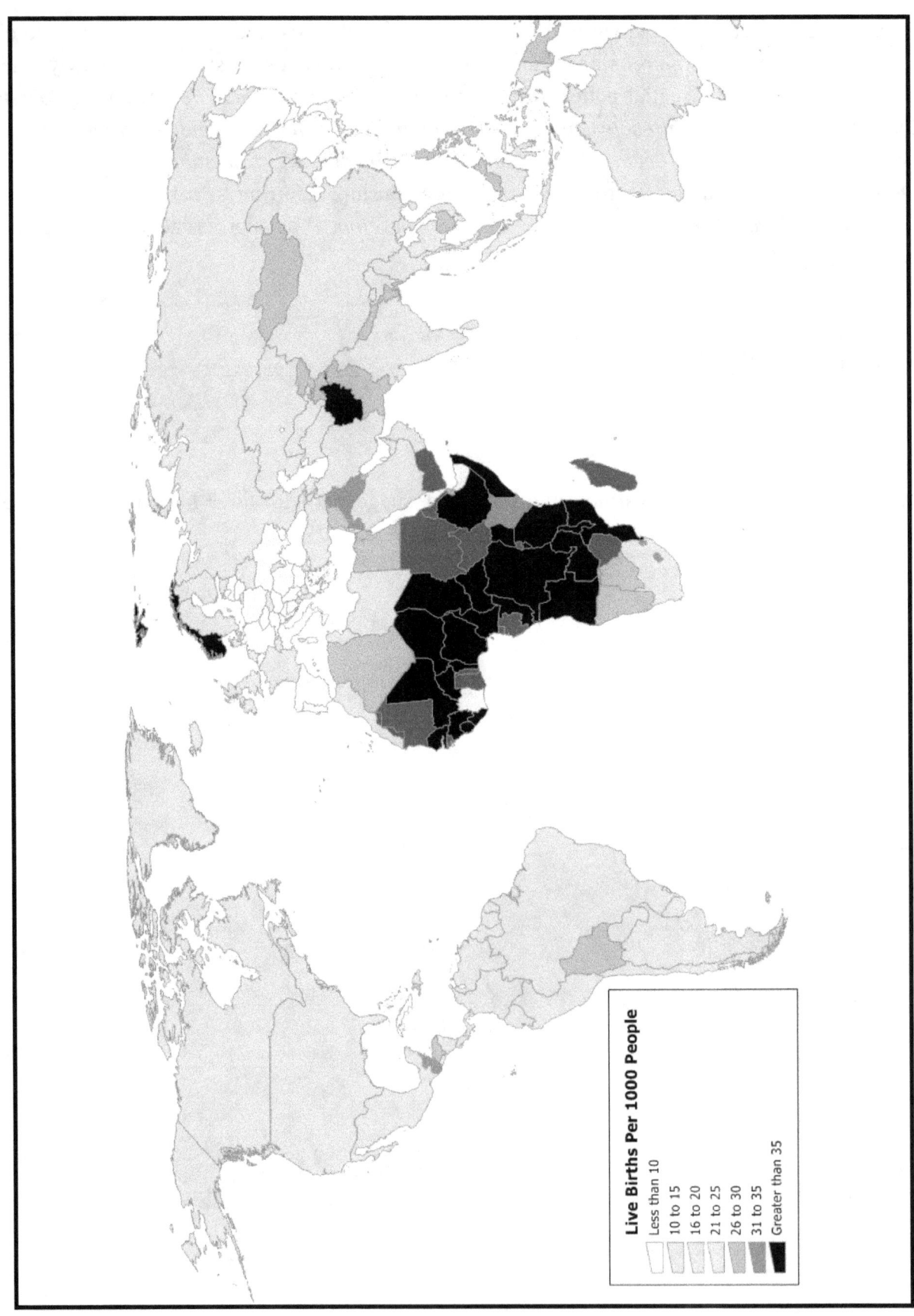

Workbook for Statistical Problem Solving in Geography

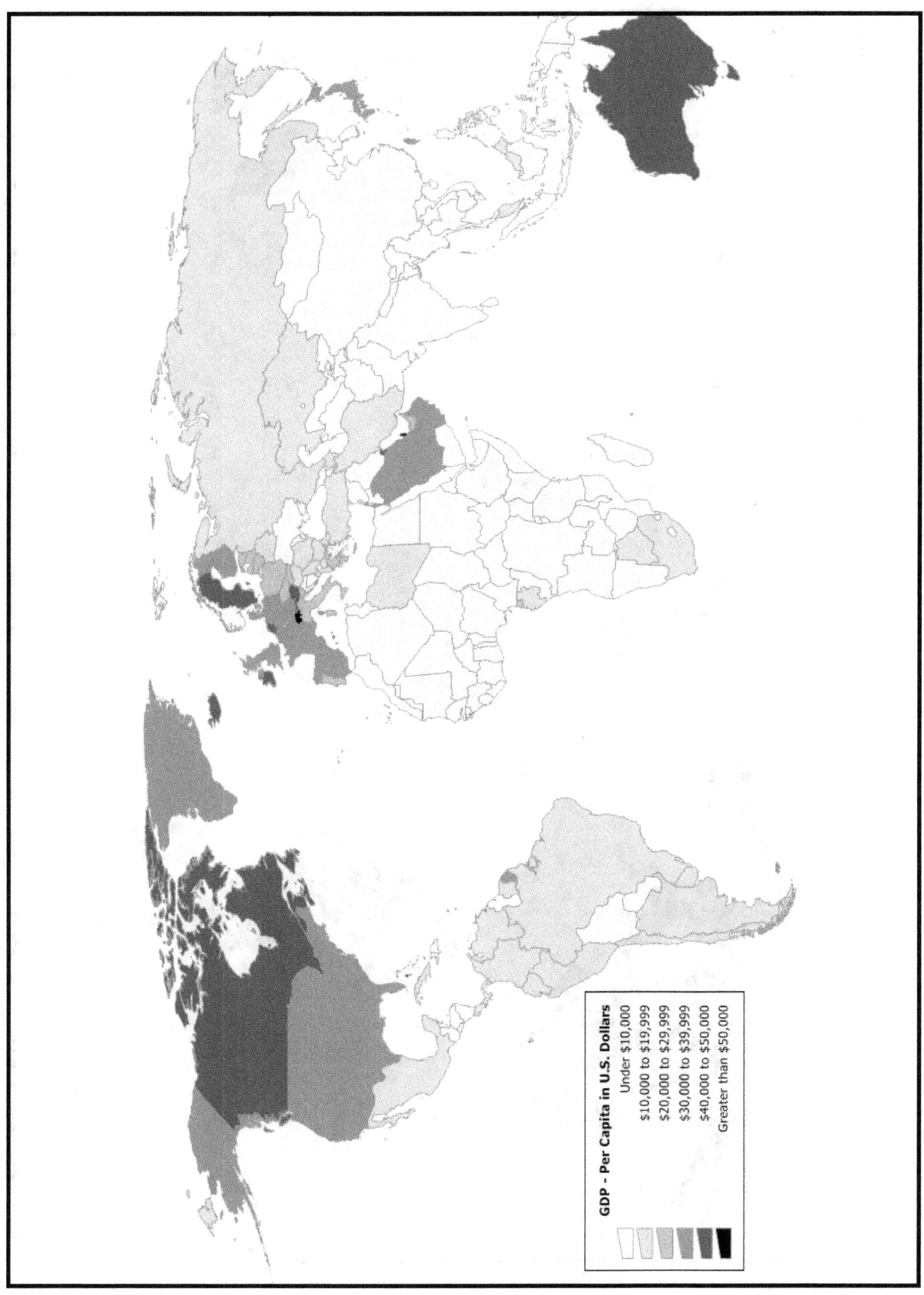

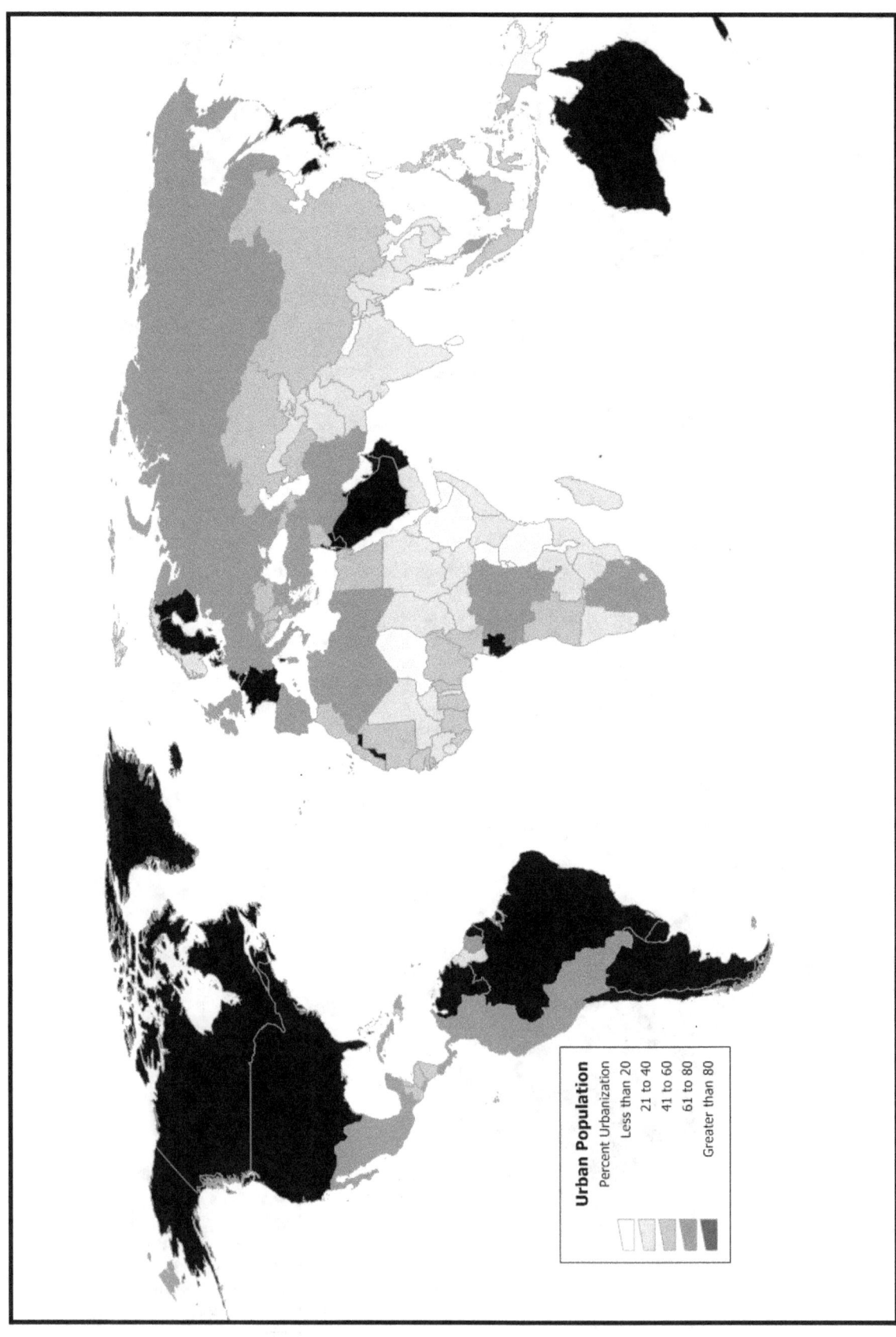

Workbook for Statistical Problem Solving in Geography

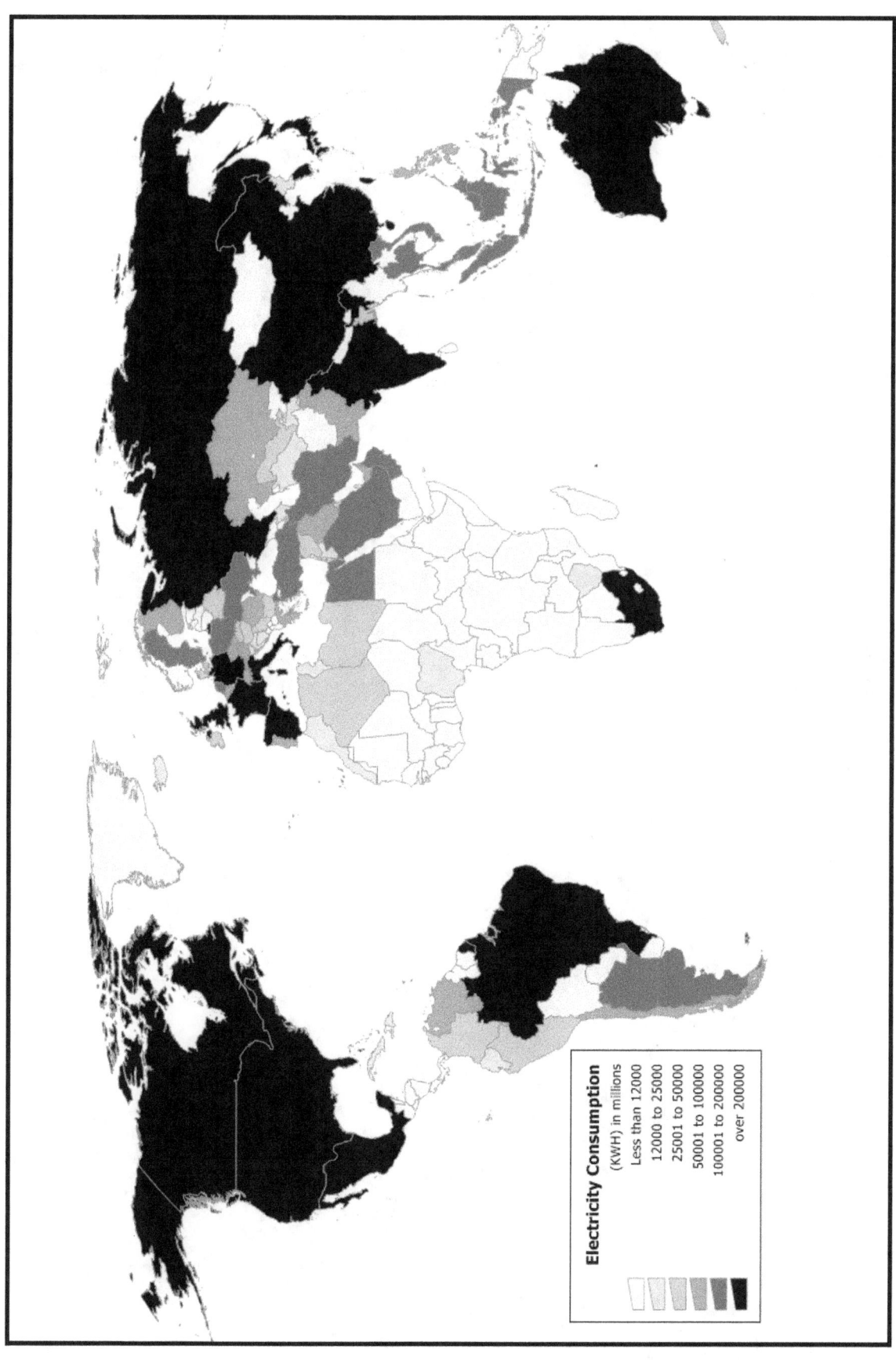

Chapter 2
Geographic Data: Characteristics and Preparation

Major Goals and Objectives in this Chapter

In chapter 2 of *An Introduction to Statistical Problem Solving in Geography* you learned how to categorize geographic variables and data sets on a variety of dimensions and also identify measurement scales of a variable. You also learned how to classify a dataset using the four primary classification techniques.

In this exercise you will describe and classify a dataset related to its many dimensions, and also thematically render the results of each classification method.

Problems and Exercises:

For this exercise, you will be able to choose from numerous data collected in the last United States Census, including (1) Educational Attainment by State; (2) Gross Domestic Product by State in 2009 and 2010; (3) Births to teenage mothers (1990, 2000, and 2009), and (4) 2010 Poverty Level by State. These data are found in the accompanying spreadsheet under the tab *Chapter 2 – US Data*.

United States State-Level Data

Select one of the variables from the United States state-level data set:

Variable selected: _____

1. Categorize this variable on the following dimensions, placing the correct letter for each line in the blank:

   ____   (a) discrete variable   (b) continuous variable

   ____   (a) primary data source   (b) secondary (archival) data source

   ____   (a) individual-level data   (b) spatially-aggregated data

   ____   (a) population data   (b) sample data

   ____   (a) quantitative variable   (b) qualitative variable

   ____   (a) explicitly spatial   (b) implicitly spatial

2. Identify the scale of measurement of data for the selected variable: _____

   (a) Nominal   (b) Weakly-ordered ordinal   (c) Strongly-ordered ordinal
   (d) Interval (e) Ratio

3. Evaluate the variable you selected in terms of potential measurement error, using the four categories below. If you have reason to question any of these measurement concepts, identify the nature of the problem(s) and explain the reasons for your uncertainty:

Precision:

Accuracy:

Validity:

Reliability:

4. Classify your data using **each** of the four simple classification methods shown in Table 2.4 of the textbook: Equal intervals based on range, equal intervals not based on range, quantile breaks, and natural breaks. For consistency, use the same number of classes for each method. Identify the class breaks for each classification as part of your legend.

5. Using the maps on the following pages, create four choropleth maps of your variable, one for each classification method, similar to Figure 2.2 in the textbook. Be sure to include a clear descriptive title and legend on each map. While it is possible to generate the classifications using GIS software, doing this exercise by hand is encouraged so that you learn *how* the methods actually work.

6. Based on the maps you created, discuss the most important characteristics (either positive or negative) of each classification method.

Equal intervals based on range:

Equal intervals not based on range:

Quantile breaks:

Natural Breaks:

Which method of classification do you believe most accurately reflects the nature of the spatial pattern? Why?

Chapter 3
Descriptive Statistics and Graphics

Major Goals and Objectives in this Chapter

In Chapter 3 of *An Introduction to Statistical Problem Solving in Geography* you learned about the basic descriptive measures of central tendency, dispersion, and relative position along with the advantages and disadvantages of both. You also learned about the advantages and disadvantages of different classification schemes and also how to construct a classification scheme for a particular set of data. Finally, you were introduced to the modifiable area unit problem (MAUP) and the influences from: (a) impact of external boundary delineation, (b) the Modifiable Areal Unit Problem (MAUP), and (c) the spatial aggregation or scale problem.

Problems and Exercises:

For this exercise, we are going to analyze US Census data over the last 85 years, broken into 40 year intervals (1930, 1970, and 2010). The value of descriptive statistics in the geographic research process becomes apparent when comparing the same geographic variable over time. The data are found in the accompanying spreadsheet under the tab *Chapter 3 – US Data*.

United States Data

Select one of the three time-series variables from the United States state-level data set. The three variables are: (a) **Total Population;** (b) **Number of Farms**; and (c) **Percent Urban Population**.

1. Compute the following descriptive statistics for each year in your data set: mean, standard deviation, variance, coefficient of variation, skewness, and kurtosis. Complete the following tables:

Variable Selected: _____

Summary table of Descriptive Statistics

Year	1930	1970	2010
Mean	_____	_____	_____
Standard Deviation	_____	_____	_____
Coefficient of variation	_____	_____	_____
Skewness	_____	_____	_____
Kurtosis	_____	_____	_____

2. Describe how the mean, standard deviation, and variance have changed over time. Pay particular attention to the fact that these are **absolute descriptive measures**, and that they are a direct function of the magnitude of the data.

3. Comparison of the **relative descriptive measures**, such as the coefficient of variation, skewness, and kurtosis, often allow meaningful geographic insights. To observe the changes over time in these relative measures for your selected variable, construct a graph of the coefficient of variation for your selected variable. Briefly explain the changes to over time to the data.

4. Using only **the most recent date (2010)** in the time series for your variable, classify the data using *five equal intervals based on range*. This grouping method provides class intervals of uniform width with easily understood class break values. It also permits calculation of weighted mean, standard deviation, variance, and coefficient of variation using the midpoint and frequencies of the class breaks (see tables 3.4, 3.8 to see how to compute the data). Compare these statistics with the unweighted values you computed earlier:

Variable selected _____

Summary Table of Descriptive Statistics		
Statistic	Weighted Value	Unweighted Value
Mean		
Variance		
Coefficient of Variation		

5. Offer possible explanations for the differences between the weighted and unweighted values in the summary table:

The Modifiable Area Unit Problem

As discussed in Section 3.4 in the textbook, the Modifiable Area Unit Problem (MAUP) presents a challenge to geographers due to the changes in descriptive statistics based upon the boundary areas. Using the information for *Land in Farms* calculate the mean, standard deviation, and coefficient of variation of the State and US Census Region summaries. Refer to Table 3.11 in the textbook for an illustration of this using the Total Population values.

6. Discuss the impact that the different levels of spatial aggregation have had on the underlying statistics.

In addition to the boundary problem, the MAUP presents a challenge when grouping data based on different boundaries. The data tab *Chapter 3 – Snowfall and Precipitation* in the accompanying spreadsheet list 98 US Cities along with their rainfall and snowfall totals. Each city is also designated with one of three different zones for both the East/West (E,C,W) orientation and North/South orientation (N,C,S). As an example, those cities designated "E" represent those locations that are in the eastern-most third longitude. Similarly, those cities designated "N" represent those cities in the northern-most third longitude.

7. For this example, compute the average and standard deviations of precipitation snowfall totals for each regional grouping. What trends do you notice in the different zoning patterns?

Grouping	Average Precipitation	Standard Deviation (Precipitation)	Average Snowfall	Standard Deviation (Snowfall)
East-West Zone				
Eastern most cities (E)				
Central most cities (C)				
Western most cities (W)				
North-South Zone				
Northern most cities (N)				
Central most cities (C)				
Southern most cities (S)				

8. Create a scattergram of the data with average precipitation as the X axis, and average snowfall as the Y axis.

Chapter 4
Descriptive Spatial Statistics

Major Goals and Objectives:

In Chapter 4 of *An Introduction to Statistical Problem Solving in Geography* you learned about the concept of central tendency within a spatial context and learned the distinctive characteristics of descriptive spatial statistics. Within the spatial context you were introduced to measures of position, dispersion, and relative position.

Problems and Exercises:

Part I. Spatial Measures of Central Tendency for Major League Baseball Team Locations

For this exercise you will explore the westward migration of Major League Baseball teams over the last 100 years. The data tab *Chapter 4 – Baseball Teams* in the accompanying spreadsheet shows the cities of Major League Baseball teams from 1900 to 2010 in twenty year intervals. As the country expanded westward, one might expect a corresponding western movement of baseball teams to serve the growing population.

1. For each grouping of years, compute the mean centers for the baseball team locations (refer to Tables 4.3 and 4.5 in your textbook for the computation).

Year	Mean Center (X,Y)
1900	_____ , _____
1920	_____ , _____
1940	_____ , _____
1960	_____ , _____
1980	_____ , _____
2000	_____ , _____
2010	_____ , _____

2. Plot the locations of the mean centers on the map provided. What do you observe about the center of gravity and general dispersion of Major League Baseball team locations over the last century? Also, how does this map compare to the geographic center of the U.S. Population shown in Figure 4.5 in your textbook?

Part II. Weighted Spatial Measures of Central Tendency for Canadian Provinces

The data tab *Chapter 4 – Population and Farms* in the accompanying spreadsheet shows the data for the Population (2013) and Number of Farms (2004) for each Province in Canada.

3. Calculate the Weighted Mean Center and Weighted Standard Distance for both the Number of Farms and Population in Canada.

	Weighted Mean Center (latitude, longitude)	Weighted Standard Distance
Population	____ , ____	_____
No. of Farms	____ , ____	_____
Unweighted	____ , ____	_____

4. Plot the unweighted Province centroids, and the Weighted Mean Centers for the 2013 Population and the 2004 Number of Farms on the map provided. What do you observe about the relationship of these variables with one another?

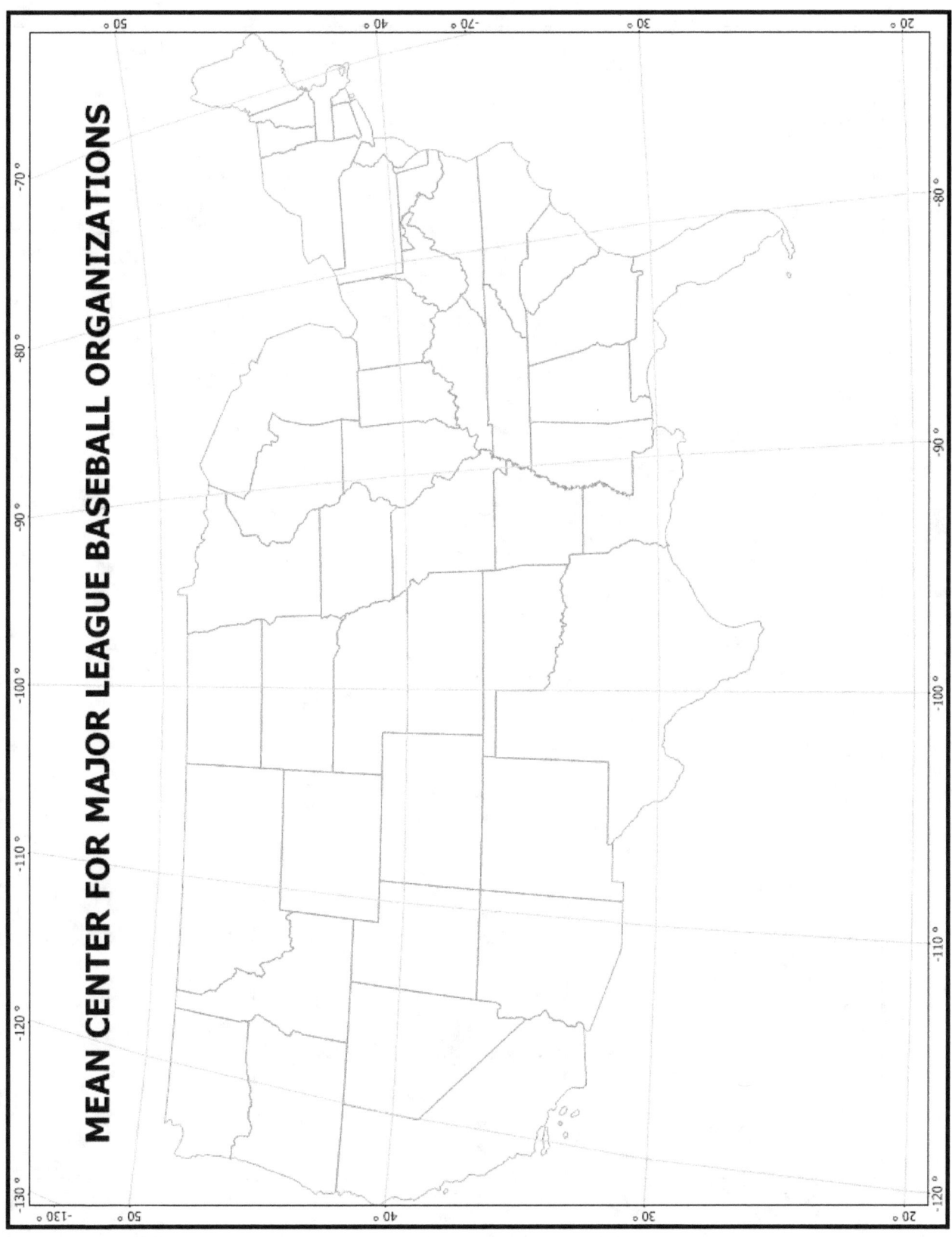

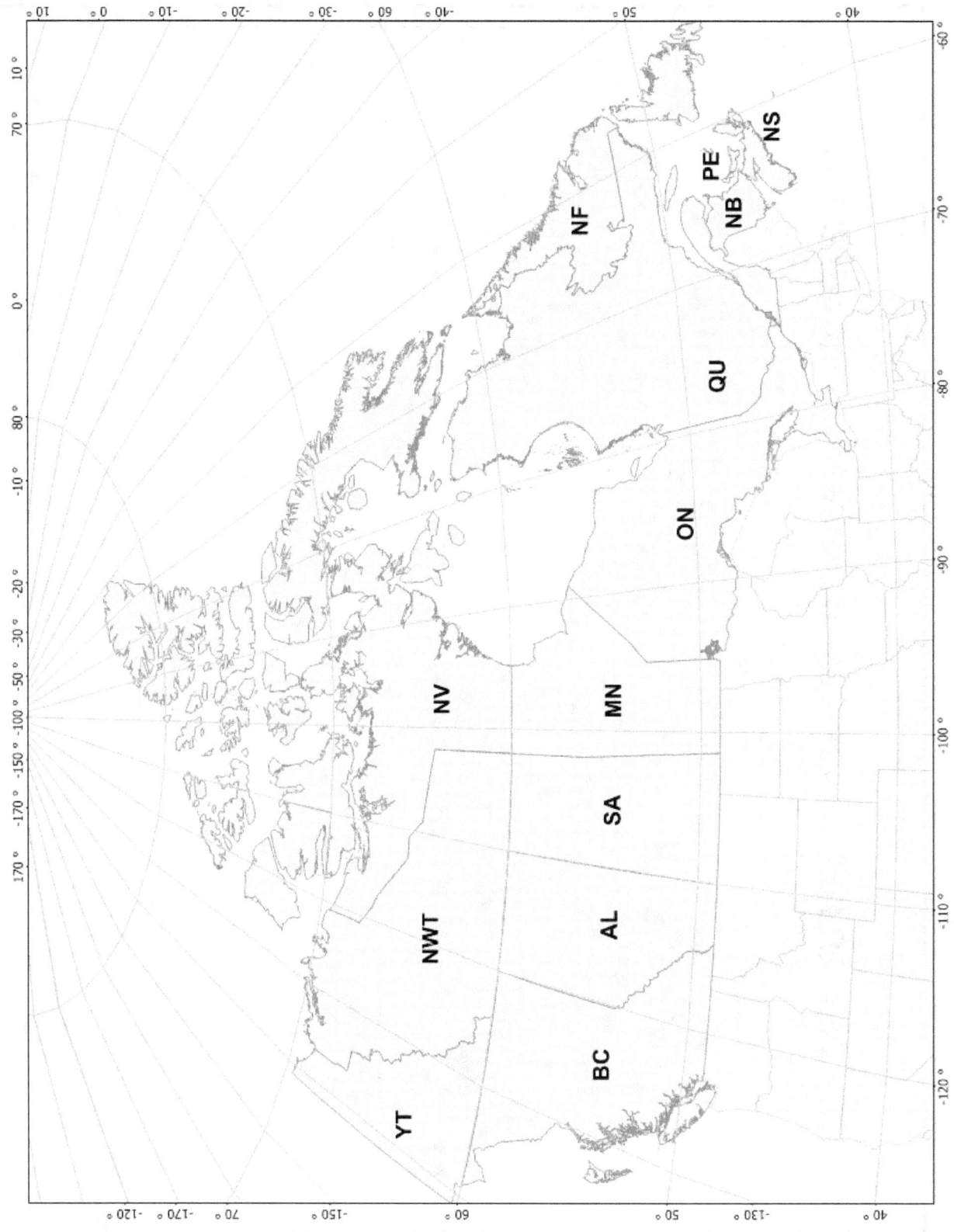

Chapter 5
Basic Probability
and
Discreet Probability Distributions

Major Goals and Objectives

In chapter 5 of *An Introduction to Statistical Problem Solving in Geography* you learned about the general nature of deterministic and probabilistic processes in geography, with a special emphasis on frequency distributions and the basic rules of probability. You also learned how to apply discreet probabilistic approaches to geographic data through the binary, geometric, and Poisson processes.

For these exercises you will explore basic probability related to United States immigration, and then focus on practical applications of the geometric, binary, and Poisson probability distributions within the context of natural disasters of hurricanes, hailstorms, and tornadoes

Problems and Exercises

Part I – Basic Probability

Problem Setting: The United States is discussing the topic of immigration on a number of different fronts: political, economic, security, humanitarianism. When evaluating the impact that immigration has on particular issues (i.e. workforce, government assistance, crime, education) it is often helpful to determine the probabilities that particular issues may have as it relates to immigration.

Refer to tab *Chapter 5 – Immigration* in the accompanying spreadsheet to answer the following questions. This data list the number of persons obtaining lawful permanent resident status by region for the largest cities in each State (note there are not 50 cities because some cities like Wilmington, DE and Newark, NJ are actually absorbed into the larger metropolitan areas of Philadelphia and New York City respectively):

1. From a list of all 990,533 persons obtaining lawful permanent resident status, what is the probability of selecting an individual that is either from Asia or Africa?

2. What is the probability of selecting a person from Asia who also settled in Jackson Mississippi? What is the probability of selecting a person from Asia who settled in Seattle, Washington? How do these probabilities differ, and why?

3. What is the probability of selecting a person from Mexico or South America who lives in Houston, Texas? How does this differ from the probability of selecting a person from Mexico or South America within the entire US Population?

Part II – Discreet Probability Distributions - Severe Storms in North Carolina

Problem Setting: The State of North Carolina is vulnerable to severe weather in the form of hurricanes and tropical depressions. Many of these severe weather events have caused extensive damage to property, human fatalities, and economic hardship. As shown in tab *Chapter 5 - N.C. Hurricanes* of the accompanying spreadsheet, there have been 75 storm events ranging from Tropical Depressions to Category 5 storms in North Carolina since 1852. Recently, Hurricane Irene caused $15.8B in damage, while the Category 5 Storm Diana (which was downgraded to a Category 2 at landfall) caused $65M. Obviously, hurricanes that make landfall in North Carolina are significant and costly events.

The Binomial Distribution - Severe Storms in North Carolina

There were 23 Category 3 or greater hurricanes to impact North Carolina between 1852 and 2011. The recent lull of no hurricane activity between 2011 and 2014 is believed by climatologists to not be terribly uncommon, as they cite similar and greater lulls between 2003-2011 (Isabel and Irene representing the only hurricanes greater than a Category 2), 1985-1996 (Kate and Bertha representing the only hurricanes greater than a Category 2). Given the current budget estimates for emergency response, officials believe that having more than 2 hurricanes over the next five years would place a significant strain on Local and State government's ability to financially respond to these events.

4. Calculate the Binomial probability of receiving 0, 1, 2, 4, or 5 Category 3 or greater hurricanes within the next 5 years. Refer to Table 5.2 in your textbook for an example computation. If you were to advise the officials about the likelihood of overwhelming the State budget with multiple hurricanes in the next 5 years, what would you conclude? **Remember, we are only interested in Category 3 or greater storms.**

Number of Years with at least one hurricane over the next five years	Binomial Probability	Type of Outcome (suitable/unsuitable)
0		
1		
2		
3		
4		
5		

The Geometric Distribution - Severe Storms in North Carolina

As of September, 2014, the last severe storm to impact North Carolina was in 2011. Given this short-term lull in hurricane activity, emergency personnel and planners want to know the likelihood that another hurricane will hit North Carolina in the next 5 years, given that there have been no hurricanes in the last 3 years.

5. Calculate the Geometric Distribution to determine the probabilities associated with the next major hurricane in North Carolina. Refer to Table 5.5 in your textbook for an example of flood frequencies in the Susquehanna Basin.

The Poisson Distribution over time - Severe Storms in North Carolina

Problem Setting: Hurricanes can be considered rare events, and while is isn't unlikely to receive a hurricane in any single year, the likelihood of receiving multiple hurricanes in a single year are quite unlikely. Nonetheless, when a State is impacted by multiple hurricanes in a single year, the damages can be financially and emotionally severe. In reviewing the tab *Chapter 5 – N.C. Hurricanes*, you can see that there were some years that North Carolina received more than one hurricane (i.e. Barry and Gabrielle in 2007; Kate and Gloria in 1985; Ione, Diane, and Connie in 1955).

6. Using the data from the previous question, calculate the Poisson probabilities of North Carolina receiving 0,1,2,3,4, and 5 hurricanes in a single season. Compare these probabilities with the actual number of hurricanes. Refer to Table 5.6 in your textbook to review the Poisson calculation.

Number of Hurricanes per Year	Observed frequency (years)	Total Hurricanes	Observed Probability of Occurrence	Poisson Probabilities	Expected Frequencies (years)
0					
1					
2					
3					
4					
5					

Workbook for Statistical Problem Solving in Geography 29

7. Compare the observed frequencies of hurricanes in North Carolina with the expected frequencies, and indicate whether hurricane occurrence in North Carolina is represented as a random Poisson process.

Poisson Distribution over Space - Hailstorms and Tornadoes in Arkansas

Problem Setting: Both hailstorms and tornadoes are naturally occurring rare events that cause substantial damage to life and property. The general rule of thumb is that when the conditions are right, hail and tornadoes can strike anywhere and at any time. Due to their rareness tornadoes and hailstorms make excellent candidates to examine whether they occur randomly over space.

8. Consider the pattern of tornado touchdowns over the State of Arkansas between 2000 and 2013. What do you observe through a simple graphical look at the data?

9. Calculate the Poisson probabilities for tornado touchdowns. What do you conclude about tornado touchdowns being a random event over space?

Number of Tornadoes per cell	Observed Frequencies of cells	Total Tornadoes	Observed Probability of Occurrence	Poisson Probability
0				
1				
2				
3				
4				
5				
6				
7				
8				
9				
10				

10. Consider the pattern of hailstorms in the State of Arkansas in 2013. What do you observe through a simple graphical look at the data?

11. Calculate the Poisson probabilities for hailstorms. What do you conclude about tornado touchdowns being a random event over space?

Number of Hailstorms per cell	Observed Frequencies of cells	Total Hailstorms	Observed Probability of Occurrence	Poisson Probability
0				
1				
2				
3				
4				
5				
6				
7				
8				
9				
10				

12. How do the tornado and hailstorm distributions differ, both graphically and statistically?

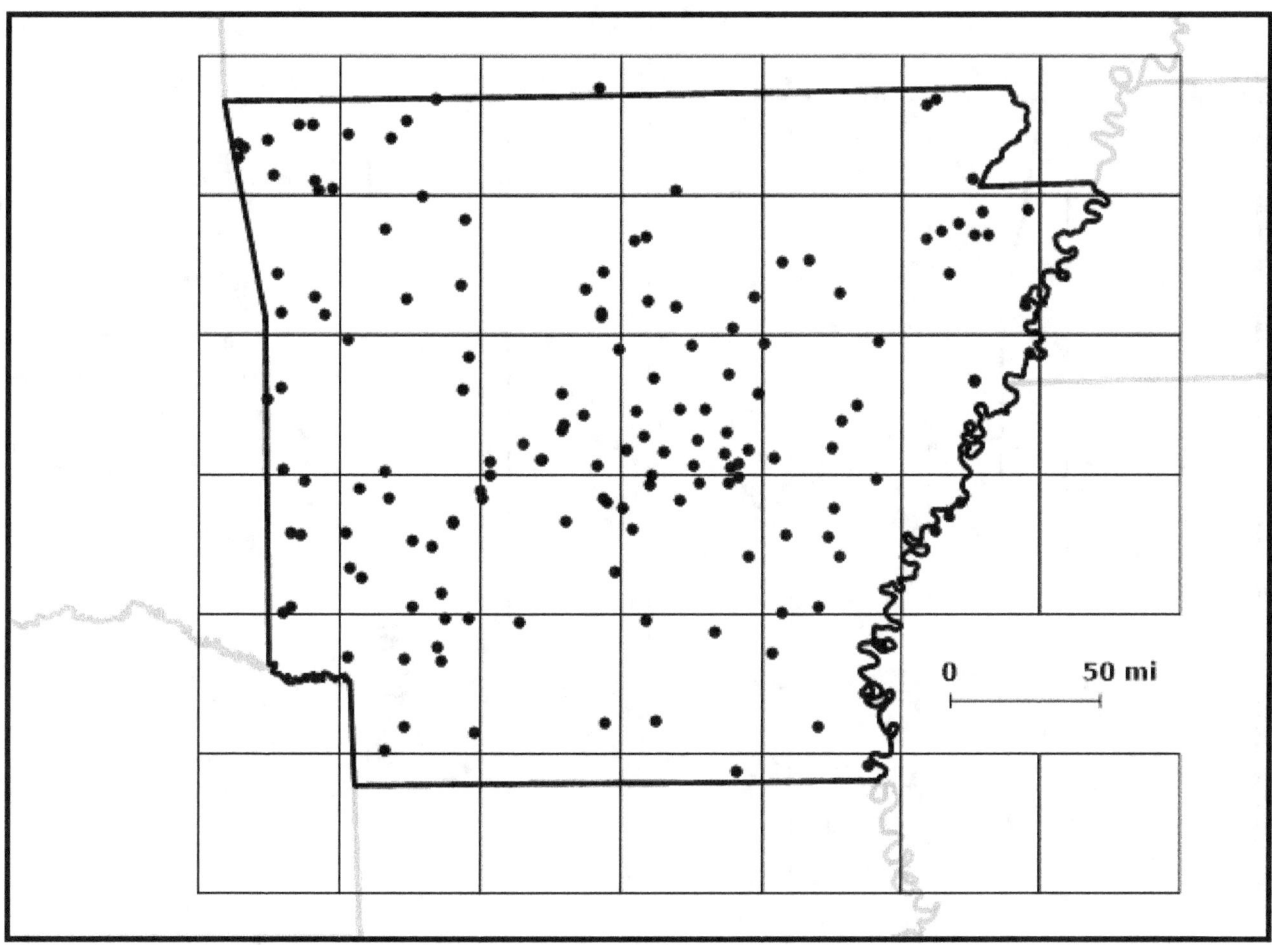

Tornado touchdowns in Arkansas

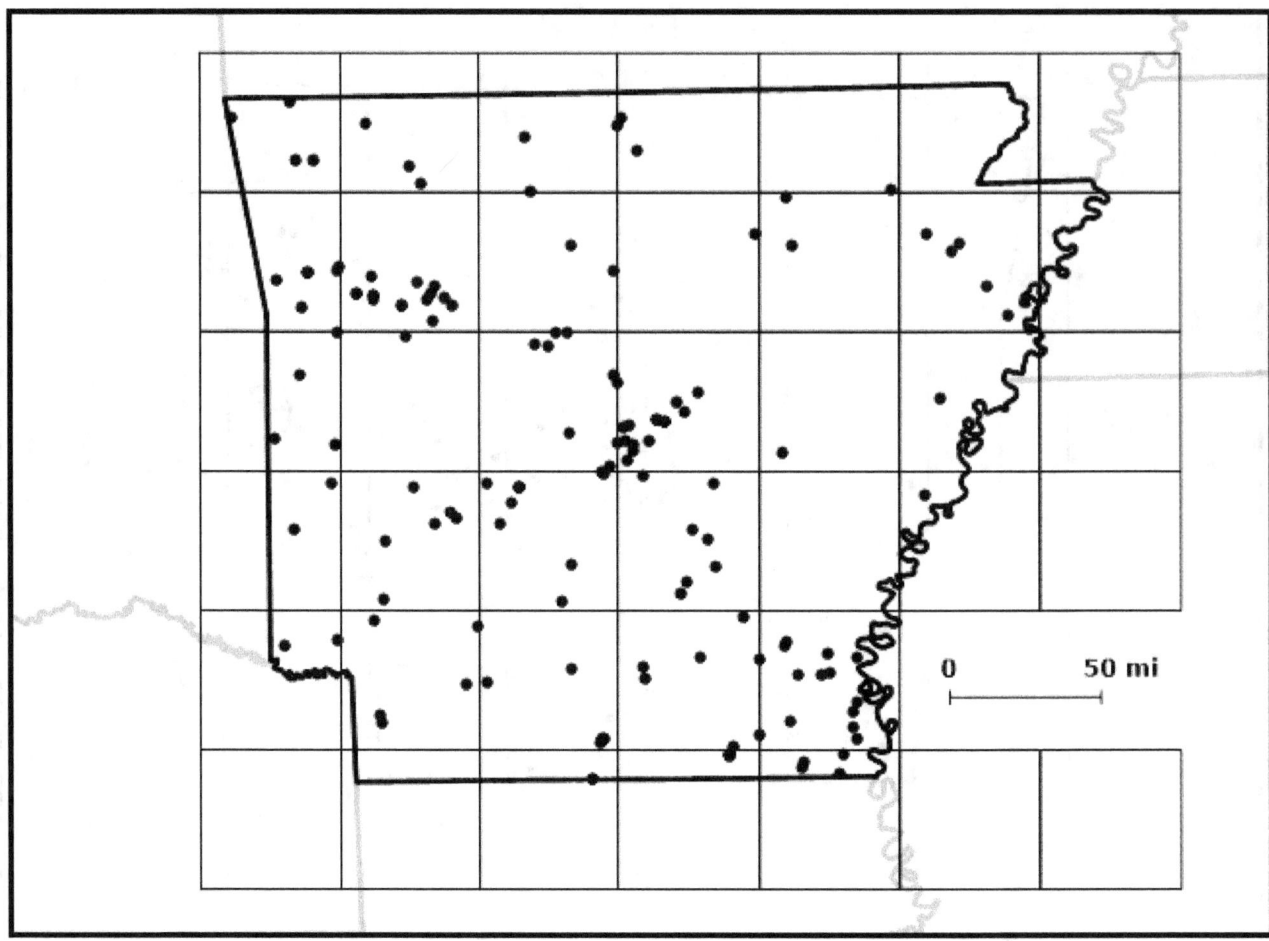

Hailstorm locations in Arkansas

Chapter 6
Continuous Probability Distributions and Probability Mapping

Major Goals and Objectives

In chapter 6 of *An Introduction to Statistical Problem Solving in Geography* you expanded your knowledge of discreet probability functions to include continuous probability functions, namely the Normal Distribution. The textbook reinforced the concept of the Normal Distribution and also demonstrated how to use a table of normal values to interpret probabilities of geographic data. Additionally, you were introduced to the concept of probability mapping.

These exercises will focus on the application of the Normal Distribution within the context of climatology, and will conclude with an application of probability mapping related to temperature for selected airports in the United States.

Problems and Examples

Part I – The Normal Distribution - Snowfall in Upstate New York

As discussed in the textbook, the Normal Distribution is the most generally applied probability distribution used in geography. In this section we will explore snowfall for three different cities in hopes of gaining a better understanding of their characteristics. Consider the yearly snowfall for Buffalo, NY, Syracuse, NY, and Central Park, NY as you answer the following questions.

City	Average Snowfall (inches)	Standard Deviation
Buffalo, NY	99.6	31.28
Syracuse, NY	131.7	44.94
Central Park, NY	33.39	19.04

1. What is the probability of each city having over 100 inches of snowfall in any given year?

2. What is the probability of each city having less than 50 inches of snowfall in any given year?

3. What is the probability of each city having between 80 and 100 inches of snowfall in any given year?

4. In 1993, Syracuse New York had over 190 inches of snow, exhausting the snow removal budget and placing a financial burden on the city. However, officials do not want to continually budget for 200 inches of snow, if the likelihood of receiving that much snowfall is minimal. The City has asked to determine the probability of receiving over 200 inches of snow in a given year. After your calculation, offer an opinion on whether the city should maintain a yearly budget to account for 200 inches of snow?

Part II Probability Mapping - Temperature in the United States

5. Construct two probability maps, one for temperatures in December, January, February (DJF) and one for June, July, August (JJA) using the means and standard deviations for the airports shown in tab *Chapter 6 - Airports* in the accompanying spreadsheet. To complete this task, determine the temperature exceeded 90% of the time for each city - refer to equations 6.3 and 6.4 (or Figure 6.6) in your textbook to calculate the expected temperature value exceeded 90% of the time for each city. Draw isolines for both probability maps.

6. What are some observations you notice about the patterns in the maps you constructed?

Workbook for Statistical Problem Solving in Geography 36

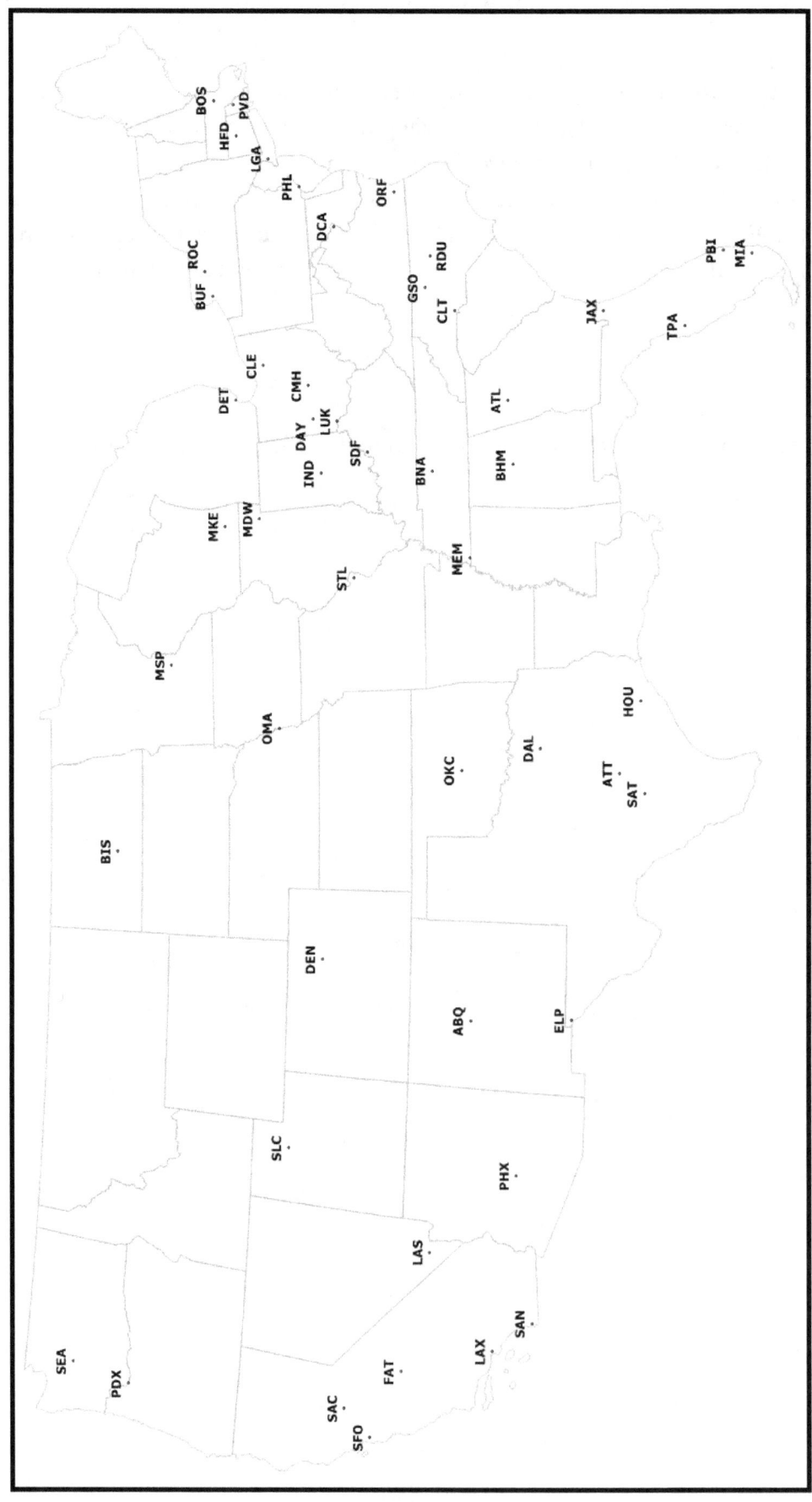

Workbook for Statistical Problem Solving in Geography

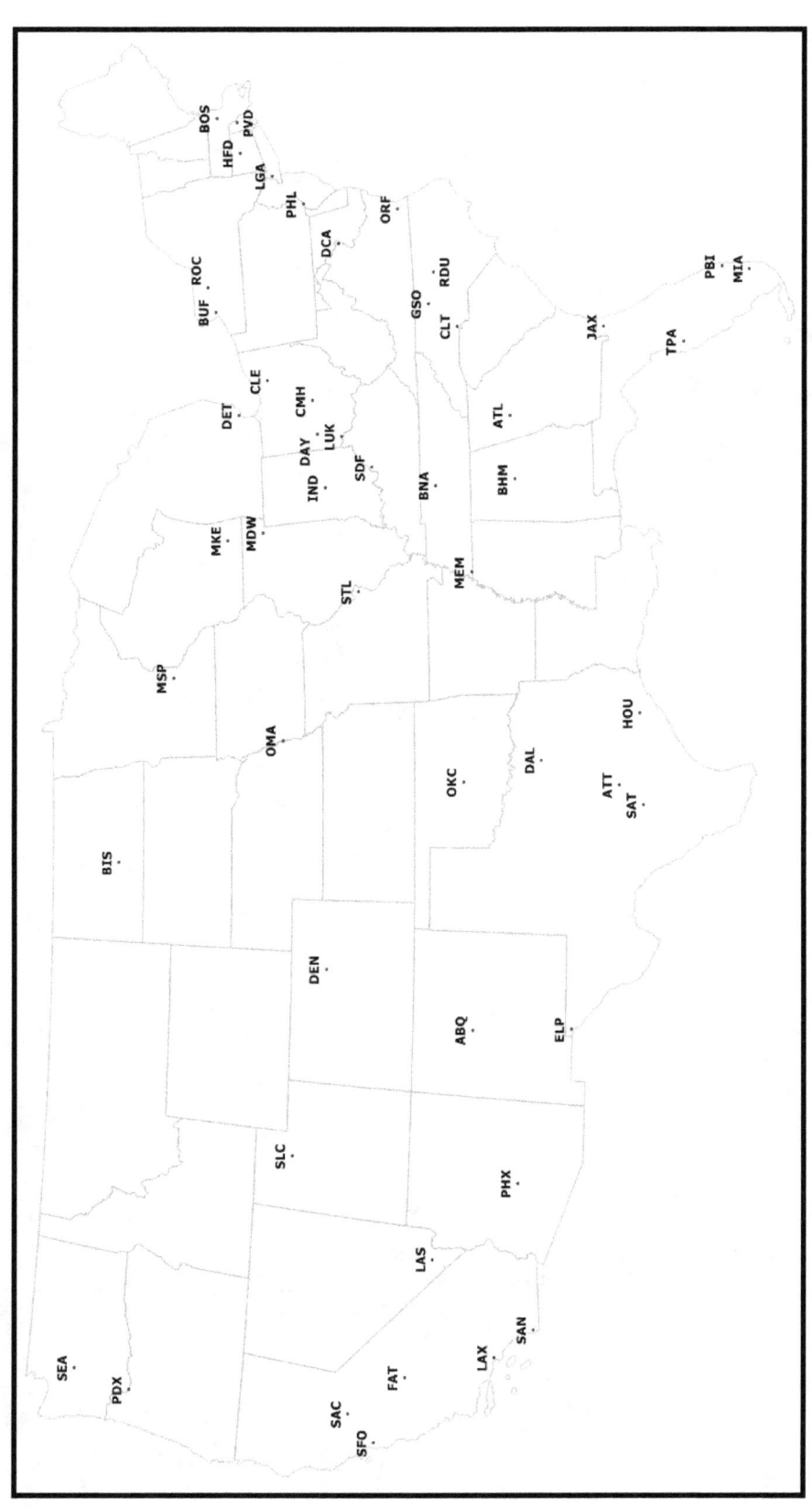

Chapter 7
Basic Elements of Sampling

Major Goals and Objectives

In chapter 7 of *An Introduction to Statistical Problem Solving in Geography* you were introduced to the advantages of sampling in contrast to a complete enumeration of the entire population. Within this context, you were introduced to basic sampling terms and characteristics of different sampling techniques. Finally, the chapter concluded with an overview of spatial sampling.

For this exercise you will be given a few geographic situations to choose from and determine the most appropriate sampling technique to employ. The exercise will conclude with an analysis of three different sampling techniques for soil data in the New York City area. For this exercise, please keep a copy of your work because you will use it in the exercise for Chapter 8.

Problems and Exercises

Part I – Sampling Designs for Geographic Situations

Listed on the next page are a number of "geographic situations", each of which requires sampling as part of the overall research process. Select two of these example problems and answer the questions that follow the list.

Geographic Situations

(a) A natural resource manager is interested in monitoring the spatial pattern of invasive species in a local forest.

(b) An urban geographer wishes to examine the locational variation of "housing quality" in a metropolitan area which includes a large central city and numerous suburban developments in several adjacent counties.

(c) A hydrologist is studying the changes in the spatial distribution and intensity of nitrogen and phosphorus levels resulting from agricultural runoff into a bay. What type of spatial point sampling procedure to monitor these activities?

(d) An urban planner wishes to survey area residents to learn their attitudes regarding various downtown revitalization strategies.

(e) A forester wants to examine environmental threats in the Everglades, including (but not limited to) Big Cypress National Preserve, and wishes to take a spatial sample from this area (see figure 6.3).

(f) An emergency planner wants to study the attitudes of residents living on the Outer Banks in North Carolina barrier islands toward alternative coastal zone management strategies. They want to understand how people's attitudes change over time.

(g) An economic geographer wants to analyze the pattern of growth in a community by assessing the attitude of residents in a neighborhood before, during, and after construction of a new skate park.

Answer the following questions for each of the "geographic situation" you selected.

Workbook for Statistical Problem Solving in Geography

First Geographic Situation (indicate letter) _____

1. Define the target population and target area for the example situation.

2. Designate the sampled population and sampled area. Clearly explain your operational definition for the sampling frame.

3. Identify the type of probability sampling procedure you would employ. Provide a concise, but sufficiently complete, description of the sampling design selected (for example, systematic aligned point sample).

4. Select an appropriate method of data collection (for example, mail questionnaire) and justify your choice. Also, indicate the operational plan you would use to sample.

Workbook for Statistical Problem Solving in Geography

Part II – Spatial Sampling of Soils in New York City:

The General Soil Map for the New York City Metropolitan area is shown with six drainage classifications – Dune Land (DL), Excessively Drained (ED), Somewhat Excessively Drained (SED), Urban Designation (UD), Very Poorly Drained (VPD), and Well Drained (WD). The following table lists the percent land area, computed with GIS software, that each classification covers.

Drainage Classification	Percent Land Area (Raw Data)	Simple Random	Simple Systematic	Systematic Unaligned
DL	0.01			
ED	0.01			
SED	0.41			
UD	0.41			
VPD	0.03			
WD	0.13			

For this data, we have the actual percent land area covered by each drainage classification - in reality, when sampling in the field, we will not have this luxury (which is why we sample in the first place). This luxury however is useful to us to learn the differences in the sampling methods and determine if one particular method better represents the actual conditions in the field.

In the column *Percent Land Area*, drainage classes SED and UD account for over 80% of the land area, while WD represents 13% and VPD represents 3%. A good spatial sample should show a similar distribution.

5. Using the following maps, count the number of points in each drainage classification, determine the percentage of points that fall into each category, and fill in the table above.

6. Indicate how well each of the sampling techniques matched the raw data.

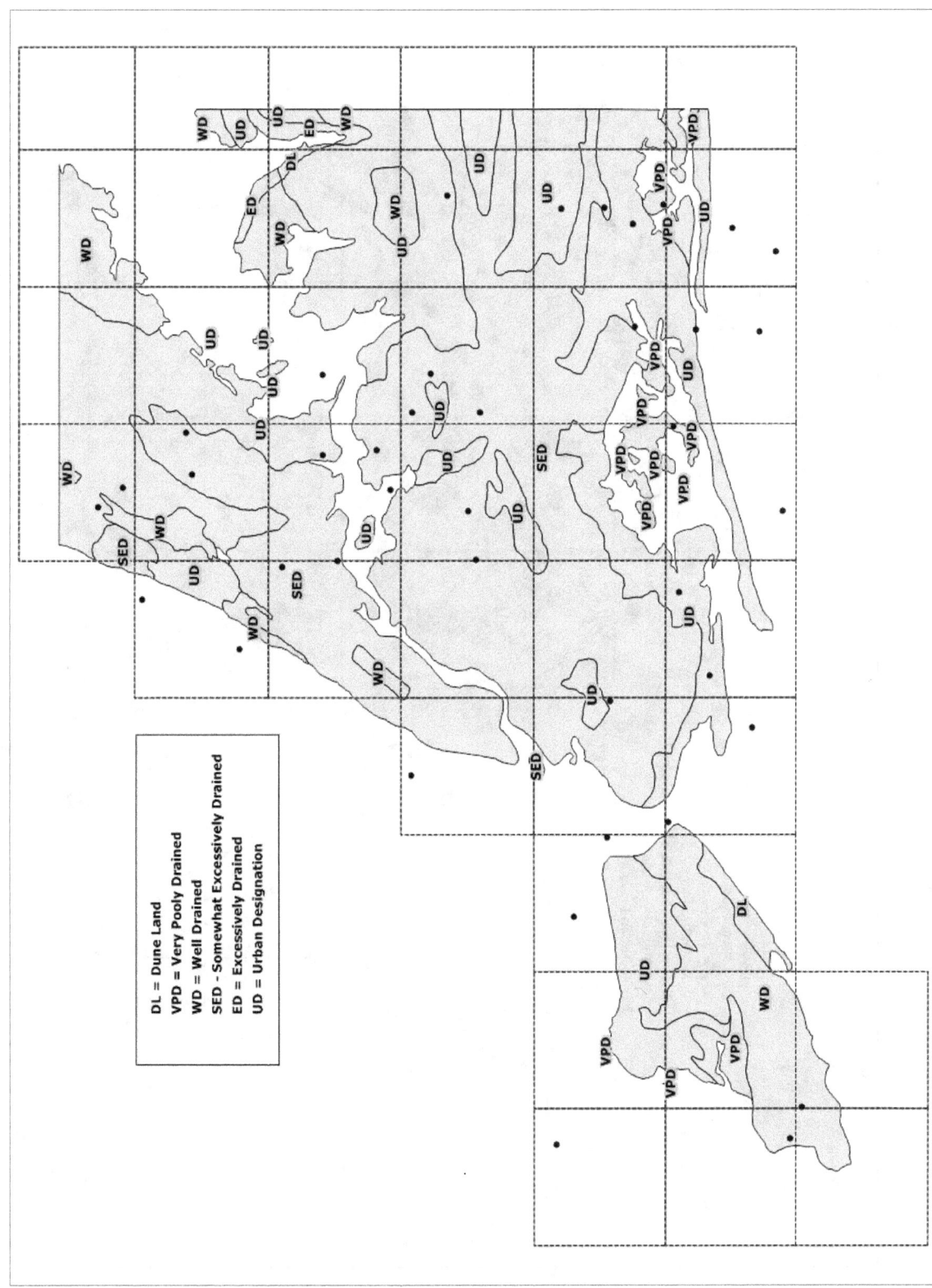

Figure 1 Simple Random Spatial Point Sample

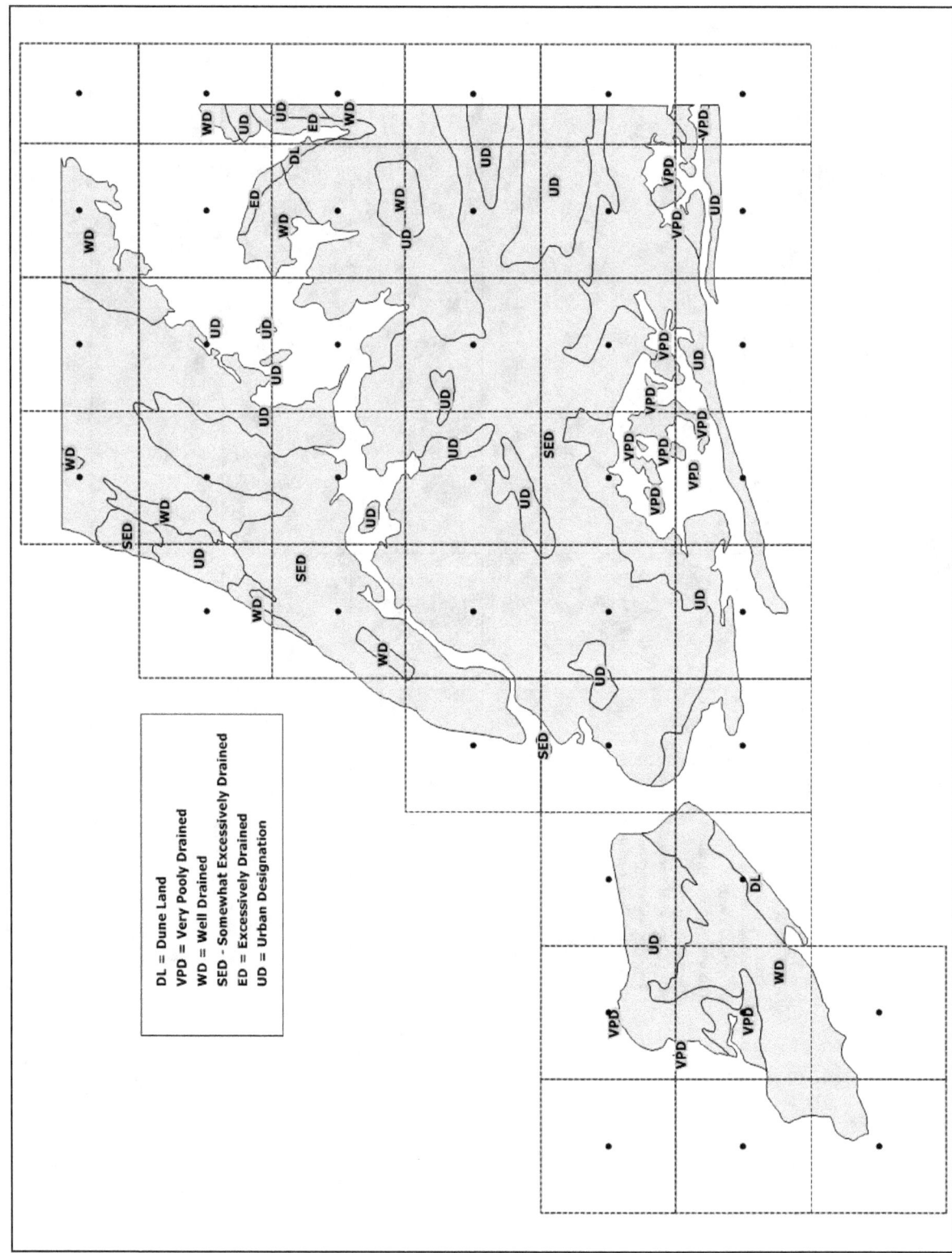

Figure 2 – Systematic Spatial Point Sample

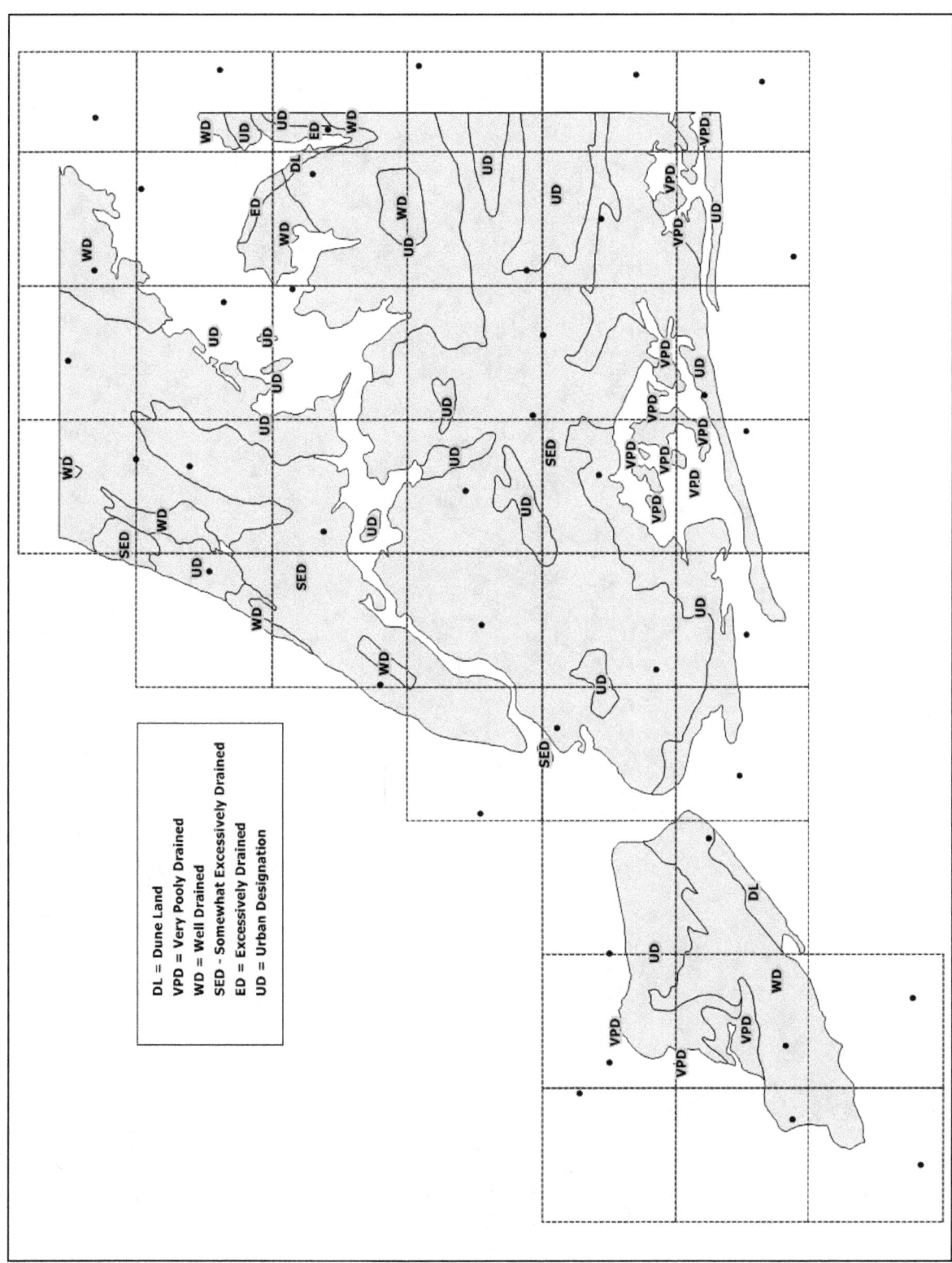

Figure 3 – Systematic Unaligned Spatial Point Sample

Chapter 8
Estimation in Sampling

Major Goals and Objectives

In chapter 7 of *An Introduction to Statistical Problem Solving in Geography* you learned about the concepts of point and interval estimation. The textbook also covered the more theoretical concept of the Central Limit Theorem, which is foundational to the application of inferential statistics. Within that context, you learned how to develop the best point estimates for a sample and also how to develop confidence intervals around those estimates.

For this exercise you will apply the Central Limit Theorem concept to a random sample of agricultural yield values and evaluate the practical application of the theory. In addition, you will develop confidence intervals for the different sampling techniques used in Chapter 7. Finally, a number of real world scenarios in the context of urban planning are introduced and confidence intervals are generated.

Problems and Exercises

<u>Part I – Confidence Interval Estimation and the Central Limit Theorem with Agronomic Data</u>

For this part, we are going to analyze agricultural yield data. Tab *Chapter 8 – Central Limit Theorem*, in the accompanying spreadsheet contains random samples from the agricultural yield example (Table 8.2) in the textbook. Fifty random samples of 40 yield values (n=40) were taken and their average and standard deviations recorded. In addition, fifty random samples of 5 yield values (n=5) were also taken and their average standard deviations recorded. Recall that when all 7,000+ values were analyzed, the original data had a mean of 123.16 and a standard deviation of 38.5.

Workbook for Statistical Problem Solving in Geography

1. Compute the mean and standard deviation for both groups of samples. Also, compute the theoretical standard error for both samples. Compare the *actual* mean and theoretical standard errors to the computed samples.

Samples	Average	Standard Deviation	Theoretical Standard Error
Original	123.16	38.5	---
50 Samples of 5 yield values			
50 samples of 40 yield values			

2. Compute a 90% confidence interval and upper and lower bounds for each of the 50 samples where n=40. Refer to Table 8.3 in your textbook for an example.

Figure 8.5 in your textbook provides an idealized example of a 90% confidence interval. We say *idealized* because it just so happens to have 90% of the confidence intervals (9 out of 10) cover the population mean with only *sample 5* falling outside of the confidence interval. The figure works well as an illustration of the concept, but real data may have slightly different results.

3. What percentage of the samples and confidence intervals computed in the previous question actually contain the population mean? Why do you think it is not exactly 90%?

Workbook for Statistical Problem Solving in Geography   49

Part II – Confidence Interval Estimation for Random and Stratified Samples

For this exercise, we return to our soil example in Chapter 7. As stated in the previous exercise, we have the luxury of already knowing the land area of each soil type, so our generation of confidence intervals will allow us to determine if our samples provide a reasonable estimate of the true proportions.

4. Enter the **best point estimate** for the proportion of soil drainage class covered by Urban Designation, Somewhat Excessively Drained, and Well drained soils (recall that the sample proportion represents the best point estimate of the population proportion).

5. Calculate the **standard error of the proportion** for both samples and enter these values as confidence intervals in the table below. Use the 95% confidence level.

Soil Association	Simple Random Sample		Simple Systematic Sample		Systematic Unaligned Sample	
	Point Estimate	Confidence Interval	Point Estimate	Confidence Interval	Point Estimate	Confidence Interval
UL						
SED						
WD						

6. Based upon the **actual values** shown in Chapter 7, which sampling design was better for constructing the confidence interval and why.

Part III – Point and Interval Estimates for Random Samples: Water Meter Readings

The data shown in tab *Chapter 8 – Water Meters* are actual water meter readings in thousands of gallons over four quarters in 2013 for a small city in Maryland. The Water Conservation Bureau wants to gain a better understanding of water usage for apartments, townhouses, and residential condominiums in the City, so they sampled 40 residents. The data include a proportional stratified sample of homes based on the dwelling type. Again, in our example we have the luxury of actually knowing the population values for water usage - so it will enable us to understand how reasonable our estimates from the 40 household survey are.

7. What is the estimated **average water usage** per household for the City. Calculate the standard error of the mean and place a 90% confidence interval around the sample mean.

Average Water Usage by Residential Type				
Residential Strata	**Population Size**	**Sample Size**	**Sample Mean**	**Sample Standard Deviation**
Apartments				
Townhouses				
Residential Condominiums				
			Best Point Estimate	**Confidence Interval**
Totals				

8. Based on information from the City water department, there are 1382 properties that are either apartments, townhouses, or residential condominiums. The organization is interested in estimating the **total water usage** from these properties. Calculate the standard error of the population and place a 90% confidence interval around the sample total.

Total Water Usage by Residential Type				
Residential Strata	**Population Size**	**Sample Size**	**Sample Mean**	**Sample Standard Deviation**
Apartments				
Townhouses				
Residential Condominiums				
			Best Point Estimate	**Confidence Interval**
Totals				

Due to pressures on the water system from drought, the Water Conservation Bureau is interested in mandating the use of energy saving devices such as low flow shower heads and reduced gallons-per-flush toilets. The Water Conservation Bureau believes that the use of these energy saving devices will not only lower water bills but also provide less stress on the City Water Department. Nonetheless, legislating the use of these devices are controversial, so the Bureau decided to ask as part of their survey whether the residents were in favor of legislation of energy saving devices.

9. Based on the responses in the survey, estimate the **proportion of residents** in favor of the proposed legislation. Calculate the standard error of the population and place a 90% confidence interval around the sample total.

Proportion of Households In Favor of Legislation of Energy Efficient Water Devices				
Residential Strata	**Population Size**	**Sample Size**	**Number Responding "Yes" to Survey**	**Number Responding "No" to Survey**
Apartments				
Townhouses				
Residential Condominiums				
			Best Point Estimate of Proportion	**Confidence Interval**
Totals				

Chapter 9
Elements of Inferential Statistics

Major Goals and Objectives

In Chapter 9 of *An Introduction to Statistical Problem Solving in Geography* you explored the concept of hypothesis testing in the form of a one-sample difference of means and one-sample difference of proportions test. The chapter introduced classical hypothesis testing in addition to the *p-value* approach for a number of geographic situations.

For these exercises you will apply the classical hypothesis approach using real data in the context of water usage and housing sales.

Problems and Exercises

Part I – One sample difference of means test - City Water Use

We will continue our exploration of the City water usage data from Chapter 8. For this question, consider the data in tab *Chapter 9 – Water Usage (Res.)*.

According to a 2005 USGS study, the average annual water usage per household in the United States is 92,000 gallons. A local non-profit environmental advocacy organization believes that water usage per residential household in their city exceeds the National Average and has begun a campaign to reduce water use in the city. Board members of the group are concerned of embarrassing publicity if they are wrong in their initial hypothesis that the City uses more water per household than the national average. The Board members want to be at least 90% confident that the water usage in the City is greater than the national average. Therefore, they sampled 40 residential homeowners and asked how much water they used in the last year. The results of the study are found in *Chapter 9 – Water Usage (Res.)*.

1. Use the classical hypothesis testing method to evaluate your hypothesis. Provide information for Steps 1-6 as illustrated in Table 9.1 in your textbook. Provide a one or two sentence explanation or rationale for each step, as appropriate:

Step 1:

Step 2:

Step 3:

Step 4:

Step 5:

Step 6:

Sketch the normal curve with relevant areas and values labeled for your problem (see Figure 9.2 in your textbook). Briefly interpret your results.

Calculate the p-value associated with this problem. Sketch a normal curve and label the information similar to Figure 9.3 in your textbook. Briefly interpret your results.

Home Sales

The US Census Bureau reported that the average selling price for new homes in 2005 was $297,000. A Realty Marketing firm believes that the average selling price of a new home in Wicomico County, MD in 2005 is less than the national average. A random sample of 40 home sales is shown in tab *Chapter 10 - Home Sales*.

2. Use the classical hypothesis testing method to evaluate your hypothesis. Provide information for Steps 1-6 as illustrated in Table 9.1 in your textbook. Provide a one or two sentence explanation or rationale for each step, as appropriate:

Step 1:

Step 2:

Step 3:

Step 4:

Step 5:

Step 6:

3. Sketch the normal curve with relevant areas and values labeled for your problem (see Figure 9.2 in your textbook). Briefly interpret your results.

4. Calculate the p-value associated with this problem. Sketch a normal curve and label the information similar to Figure 9.3 in your textbook. Briefly interpret your results.

Part II – One Sample Difference of Proportions Test Housing Prices

The housing crises of 2008 caused economic hardship for many homeowners, and nearly collapsed the US economy. According to RealtyTrac's U.S. Home Equity & Underwater Report for the first quarter of 2014, 9.1 million U.S. residential properties (**19% of all homes**) were seriously underwater. To be seriously underwater, the combined loan amount secured by the property is at least 25% higher than the property's market value. For various reasons, a coastal community believes that the housing crisis had a much larger effect on owners of town houses, and suspect that town homes have a higher rate of being seriously underwater in their mortgage difficulty. To test their concerns, 75 townhouses were surveyed for their total assessed value and current mortgage value. Of the 75 townhouses, 13 were determined seriously underwater.

5. Use the classical hypothesis testing method to evaluate your hypothesis. Provide information for Steps 1-6 as illustrated in Table 9.1 in your textbook. Provide a one or two sentence explanation or rationale for each step, as appropriate:

Step 1:

Step 2:

Step 3:

Step 4:

Step 5:

Step 6:

6. Sketch the normal curve with relevant areas and values labeled for your problem (see Figure 9.2 in your textbook). Briefly interpret your results.

7. Calculate the p-value associated with this problem. Sketch a normal curve and label the information similar to Figure 9.3 in your textbook. Briefly interpret your results.

8. To be seriously underwater, the combined loan amount secured by the property is at least 25% higher than the property's market value.

Chapter 10
Two Sample and Dependent-Sample (Matched-Pairs) Difference Tests

Major Goals and Objectives

In Chapter 10 of *An Introduction to Statistical Problem Solving in Geography* you extended the general concept of inferential problem solving to include the two-sample and dependent-sample difference tests. In this exercise you will perform inferential tests using two samples in the context of housing prices, water usage, and bay restoration.

Problems and Exercises

Part I – Two sample difference of means test - US Housing Collapse

1. The housing collapse of 2007/8 had serious impacts on the U.S. economy. Many people were concerned that even after many years, housing prices did not recover. Tab *Chapter 10 - Home Sales*, contain two random samples of 40 single family residential new home sales in Wicomico County, MD in 2005 and 2010. In assessing the recovery of the county, compare the two random samples and indicate your conclusion.

a. State the null and alternate hypothesis:

b. Calculate the two sample difference test to determine whether the new home sales are statistically similar or come from two different underlying populations.

c. Offer your conclusion as to whether the county has recovered from the housing collapse in 2007/8.

2. The City Council is considering legislation to reduce water usage in the City. One of the council members believes that homes with pools use more water than homes without pools, and would like to specifically add a tax levy to homeowners with pools. Two random samples of 30 homeowners with pools and 30 homeowners without pools were taken and their respective water usage was recorded in tab *Chapter 10 - Water Use*. You have been asked by the Council to assess the situation and make recommendations. Your task is not to get into the politics of whether a levy should be weighed, but to provide an expert opinion as to whether homes with pools consume more water than homes without pools.

a. State the null and alternate hypothesis:

b. Calculate the two sample difference test to determine whether homes with pools utilize more water than homes without pools.

c. Offer your conclusion as to whether the water usage varies by whether a home has a pool or not.

Workbook for Statistical Problem Solving in Geography

Farming Practices in Brazil

3. Farming practices in Brazil were compared to determine if Brazilian farmers practicing sustainable agriculture maintain greater amounts of forest land on their properties. Twenty-five random samples from farmers practicing sustainable farming and farmers not practicing sustainable farming were collected and shown in tab *Chapter 10 - Brazil Farming*. The data shows the amount of acres in forest at each farm. Researchers believe that those practicing sustainable farming should contain more forest land.

a. State the null and alternate hypothesis:

b. Calculate the two sample difference test to determine whether sustainable farming practices preserve more forest land than those farmers that do not practice sustainable farming.

c. Offer your conclusion as to whether the sustainable farming practices preserve more forest land than non-sustainable farming.

Part II – Two Sample Difference of Proportions Test - City Annexation

Concerns over leaching of harmful bacteria into the bay due to residential septic systems has caused the City Council to consider annexing a portion of the county adjacent to the city into the wastewater treatment plant. Doing so will most likely reduce the polluting of the bay, but will also require significant tax increases to current city homeowners to connect the county homeowners into the wastewater treatment plant. In addition, county homeowners annexed into the city would be required to pay a yearly waste treatment tax. Before introducing the controversial legislation, a study was conducted to compare a proportion of residents within the city and within the annexation area to determine their favorability for annexation.

4. A random survey of 95 city residents showed that 69 residents favor the annexation, while only 38 residents of 105 county residents favored annexation. Test the hypothesis that the proportion of city residents favor annexation is greater than the county residents.

a. State the null and alternate hypothesis:

b. Calculate the two sample difference of proportions to determine whether the proportion of city residents favor annexation more than county residents, and state your conclusion.

Part III – Two Sample Matched Pairs Test - Water Use

5. The City Council believes that more water is used during the summer months than during the winter months. If true, the city can expect to have higher demands in the summer, and during excessive drought, place a difficult burden on the City Water Department. A random sample of 25 homes was taken and the water usage (in 1000s of gallons) was compared for each homeowner during the winter (January, February, March) and summer (July, August, September). The survey results are shown in tab *Chapter 10 - matched pairs*.

a. State the null and alternate hypothesis:

b. Calculate the two sample matched pairs test to determine whether water usage differs in the summer vs. the winter.

c. Offer your conclusion as to whether water usage differs in the summer vs. the winter.

Dissolved Oxygen – Winter and Fall

Just like humans, fish and other aquatic species need oxygen to survive. The amount of oxygen available to aquatic species is measured in the dissolved oxygen (DO) content in the water in mg/L. Species in the Chesapeake Bay typically require dissolved oxygen concentrations of 5.0 mg/L or more to live and thrive, while areas with less than .2 mg/L of dissolved oxygen are called anoxic, and are often called "dead zones." Most areas in the Chesapeake Bay that have low dissolved oxygen levels are the result of a complex interaction of several natural and man-made factors including water temperature (warmer water typically has less dissolved oxygen) and nutrient pollution.

Tab *Chapter 10 - DO in Bay*, includes two data sets: dissolved oxygen concentrations in the Chesapeake Bay for the months of September and December, and the maximum recorded dissolved oxygen concentrations for stations during a month that received only .84" of rainfall (June 2014) and during a month that received 3.52" of rainfall (June 2013).

Theoretically, warmer water temperatures should have less dissolved oxygen than colder water as lower temperatures should have higher dissolved oxygen concentration. The first example includes 36 paired water samples in the Chesapeake Bay for the months of September and December.

6. Perform a matched-pairs t-test to determine if the colder month of December has higher DO concentration than the warmer month of September.

a. State the null and alternate hypothesis:

b. Calculate the two sample matched pairs test to determine whether the concentration of dissolved oxygen in the Chesapeake Bay differs between September and October.

c. Offer your conclusion as to whether DO concentration differs due to the season.

Dissolved Oxygen – Wet vs. Dry Months

Theoretically, during dry seasons, water levels often decrease and the flow of contributing streams and rivers into the Bay decreases. Due to the reduced flow, the water mixes less with the air, resulting in lower DO concentration. During rainy seasons, oxygen concentrations tend to be higher because the rain interacts with oxygen in the air as it falls. The second data set in tab *Chapter 10 – DO in Bay*, includes paired samples of the maximum DO concentrations measured during June of 2014 (a wet month), and June of 2013 (a dry month).

7. Perform a matched pairs t-test to determine if the effect of rainfall in the Chesapeake Bay increased the DO concentration.

a. State the null and alternate hypothesis:

b. Calculate the two sample matched pairs test to determine whether the concentration of dissolved oxygen in the Chesapeake Bay differs based on the rainfall in a given month.

c. Offer your conclusion as to whether DO concentration is effected by rainfall.

Chapter 11
Three or More Sample Difference Tests (ANOVA)

Goals and Objectives

In Chapter 11 of *An Introduction to Statistical Problem Solving in Geography* you learned the concept of two-sample difference tests to include three or more samples, otherwise known as Analysis of Variance (ANOVA). In this exercise you will perform inferential tests using three or more samples in the context of environmental protection of the Chesapeake Bay, and air quality in the United States with specific emphasis on California.

Problems and Exercises

<u>The Chesapeake Bay and Dissolved Phosphorous by Year</u>

The Chesapeake Bay is the largest estuary in the United States and home to a large number of plants, animals and people. Dwindling harvest of crabs and other species due to pollution runoff have concerned communities for years. One of the leading pollutants of the bay is phosphorous. Elevated phosphorus levels from fertilizers cause more algae to grow, blocking out sunlight and reducing oxygen for fish, crabs and other Bay life.

The Chesapeake Bay Program is a regional partnership that leads and directs Chesapeake Bay restoration and protection. The program also maintains an active monitoring program from over 40 agencies and institutions and house a data hub at **http://www.chesapeakebay.net/data**. The data hub contains a wealth of information related to Bay health. While we are only going to look at one particular parameter for water quality, you may wish to explore the data hub on your own and explore some of the interesting characteristics of the Bay both within certain areas and over different time frames.

Many years ago, excessive phosphorous loading in the Chesapeake Bay caused the crab harvest to reach dangerously low levels. In response, efforts to reduce phosphorous were undertaken. The following table shows the total dissolved phosphorous (TDL) in mg/L over three years (2012, 2013, 2014) in the lower Chesapeake Bay. Based on the average TDP, the amount of phosphorous in the Bay has actually decreased. However, based on the standard deviations, the difference among the years may be due to natural variation in the data.

1. Perform an ANOVA test to determine if the amount of phosphorous in the lower portion of the Bay has decreased in a statistically significant way over the last three years.

N	AVERAGE Total Dissolved Phosphorous (TDP)	Standard Deviation TDP	YEAR
183	0.012	0.0069	2012
749	0.011	0.0084	2013
296	0.005	0.0022	2014

a. State the null and alternate hypothesis:

b. Calculate the ANOVA to determine whether there are statistically significant differences among the three years in total daily phosphorous loading.

c. Write a summary conclusion of your results.

The Chesapeake Bay and Dissolved Phosphorous by Area

The following table shows the total dissolved phosphorous (TDL) in mg/L in the month of June for Chesapeake Bay, and its three largest watersheds. Knowing whether on area of the Bay has higher phosphorous loading can help stakeholders target areas for further monitoring and improvement.

2. Perform an ANOVA to determine if the amount of Phosphorous loading in the Chesapeake differs by the local watershed.

N	AVERAGE Total Dissolved Phosphorous (TDP) in MG/L	Standard Deviation TDP	WATERSHED
107	0.011	0.0041	CHESAPEAKE BAY - LOWER
113	0.025	0.0172	CHESAPEAKE BAY - LOWER MIDDLE
14	0.021	0.0083	CHESAPEAKE BAY - UPPER
16	0.016	0.0062	CHESAPEAKE BAY - UPPER MIDDLE

a. State the null and alternate hypothesis:

b. Calculate the ANOVA to determine whether there are statistically significant differences among the four major contributing watershed areas of the Bay in total daily phosphorous loading.

c. Write a summary conclusion of your results.

Ozone Concentration in the Los Angeles Basin

Ozone is a gas that can be found both at the upper levels of the atmosphere and at ground levels. While protecting us from harmful radiation at higher elevations, ground level ozone is a pollutant that contributes to smog in cities, leading to several health problems. The City of Los Angeles is frequently cited as one of the worst areas of ozone pollution in the United States.

Some believe that the higher levels of ozone in Los Angeles are due to both the natural effects occurring in the Los Angeles basin, coupled with an increase in pollution caused by the morning commute to work by millions of vehicles. Tab *Chapter 11 LA Ozone Levels*, lists the ozone levels recorded at a monitoring station in Los Angeles for 30 days in the month of July during the morning commute (7:00AM), the mid afternoon before the afternoon commute (3:00PM), and in the evening after rush hour has ended (9:00PM) - *although, some would argue that rush hour in the Los Angeles basin never truly ends!*

3. Perform an ANOVA analysis on the ozone concentration for the Los Angeles Basin.

a. Before computing any statistical analysis, indicate when you think the ozone levels will be at their highest level and why.

b. State the null and alternate hypothesis:

c. Calculate the ANOVA to determine whether there are statistically significant differences among the ozone levels during the three different timeframes for the month of July, 2014.

d. Write a summary conclusion of your results.

Kruskal Wallace - Air Quality Index

An Air Quality Index is a numerical index used for reporting severity of air pollution levels to the public. The AQI incorporates five criteria pollutants -- ozone, particulate matter, carbon monoxide, sulfur dioxide and nitrogen dioxide -- into a single index. AQI levels range from 0 (Good air quality) to 500 (Hazardous air quality). The higher the index, the higher the level of pollutants and the greater the likelihood of health effects.

The AirData website (http://www.epa.gov/airquality/airdata/index.html) provides access to air quality data collected at outdoor monitors across the United States, Puerto Rico, and the U. S. Virgin Islands. The data comes primarily from the AQS (Air Quality System) database. While we are only going to explore a few select cities, you can also use the website to obtain data for your own city.

Tab *Chapter 11 - Air Quality Index*, contains the AQI for 31 consecutive days in July of 2014 for Barstow, CA (and urban area in San Bernardino County), Boulder, CO, Chicago, IL, and New York City, NY. We want to determine if there are any statistically significant differences in air quality among these cities.

4. Because we are working with an index value, perform a non-parametric Kruskal Wallace test. Before conducting your analysis, indicate what you anticipate finding based on your understanding of these cities:

a. State the null and alternate hypothesis:

b. Calculate the ANOVA to determine whether there are statistically significant differences among the four major contributing watershed areas of the Bay in total daily phosphorous loading.

c. Write a summary conclusion of your results.

d. If you anticipated that Boulder, CO would have a more pristine air quality, were you surprised by the findings? Read the following article to understand what might be occurring: http://www.nps.gov/romo/naturescience/airquality.htm

Kruskal Wallace Chesapeake Bay Health

The University of Maryland Center for Environmental Science (UMCES) maintains scorecards for the Chesapeake Bay and other waterways in Maryland. You can explore the data at the following website: (http://ian.umces.edu/ecocheck/report-cards/). Convert the grades of A-F to numeric values of 0-4 like you see in your own GPA grades for the upper, mid, and lower Bay using the data presented below. Because the values are ordinal in nature, we need to conduct a Kruskal Wallis Test to determine if there are differences in water quality in the Bay.

For the following table, determine if the overall Bay health has improved since 2009.

Bay	2009	2010	2011	2012	2013
Mid Bay	C	C-	D	C	D+
Upper Bay	C+	C+	C	C	C
Lower Bay	C	C	C	B-	B-

5. Perform a Kruskal Wallace test to determine if the individual portions of the Bay have exhibitied similar health over the last 5 years.

Chapter 12
Categorical Difference Tests

In Chapter 12 of *An Introduction to Statistical Problem Solving in Geography* you learned about categorical difference tests such as the chi-square goodness of fit and contingency analysis. Also, you were introduced to the Kolmogorov-Smirnov test for normality. In this chapter you will explore uniform and proportional categorical differences and also assess a data set for normality.

Chi-Square Goodness of Fit

According to NOAA's National Climatic Data Center, pollen grains can accumulate in sediments and provide a record of past vegetation. Different types of pollen in sediments reflect the vegetation that was present at that time and the climate conditions favorable for that vegetation. Certain species of vegetation are more favorable to specific climates, and pollen found at a greater depth indicate the kind of climate that was in operation in the past. For example, greater amounts of pollen from deciduous trees at lower depths would indicate that the conditions supporting deciduous trees were favorable in the past. However, when less pollen is found at shallower depths, it might indicate that the change in climate and condition are no longer favorable to deciduous forests. The following data was modified from a dataset provided by Dr. Brent Zaprowski of Salisbury University.

1. Assuming a uniform distribution of pollen for the individual species, conduct a chi-square analysis to determine if the amount of pollen found at each depth, indicating a consistent climate over the years. Report on your results.

	Pollen Count at Depth Interval (cm)			
Species	**0-50**	**51-100**	**100+**	**Totals**
Conifer	303	180	180	*663*
Deciduous	336	351	516	*1203*
Grass	279	293	505	*1077*
Weeds	87	13	15	*115*
Totals	*1005*	*837*	*1216*	*3058*

Goodness of fit – distance decay

A growing religious organization was looking to expand its congregation to a new area of the region. One question the Board members were concerned about was how far people would travel before distance became an issue. A standard distance decay model indicates that interaction falls off on the order of $\frac{1}{d^2}$. A random sample of 120 visitors and the distance they traveled are presented below:

0-9.99 miles	10-19.99 miles	20-29.99 miles	30-59.99 miles
90	19	8	3

2. Using the class midpoints (5 would be the midpoint of 0-9.99 miles, 15 would be the midpoint of 10-19.99 miles, etc.) determine the expected number of visitors from a sample of 120 people based on the inverse distance formula. As an example, the number of people expected between 0-9.99 miles is computed as:

$$\frac{\frac{1}{d_i^2}}{\sum\left(\frac{1}{d_i^2}\right)} * 120 = \frac{\frac{1}{5^2}}{\sum\left(\frac{1}{5^2} + \frac{1}{15^2} + \frac{1}{25^2} + \frac{1}{45^2}\right)} * 120$$

$$= \frac{.04}{.0465} * 120 = .859 * 120 = 103$$

3. Perform a chi-square analysis of the frequencies, similar to Table 12.4 in your textbook to determine if the visitor attendance fits a distance decay model. Report on your results.

Workbook for Statistical Problem Solving in Geography

Kolmogorov-Smirnov Test for Normality - Agricultural Yield Data

In chapter 8 of *An Introduction to Statistical Problem Solving in Geography* you were introduced to an agricultural yield dataset that appeared to be positively skewed. However, the textbook never conducted a test to determine if the data was in fact normally distributed or not.

4. In this example, you will examine a random sample of 39 yield values from that dataset found in tab *Chapter 12 Yield (Normality)* of the accompanying workbook to determine if the yield values are normally distributed. As indicated in the textbook, the Kolmogorov-Smirnow test if rather cumbersome. Therefore, feel free to utilize statistical software such as Minitab, SAS, or R, or utilize an online service.

a. What is the K-S value?

b. What do you conclude about the normality of the data?

Contingency Tables - Land Cover Change

A large 2400 hectare parcel of land in rural New York was evaluated for land cover changes. Land cover maps were constructed using photographs from 1970 and 2000. Geographers interpreted the land cover into three categories: agriculture, forest, and developed land. A random sample of 120 points was collected to determine the changes in land cover from 1970 and 2000, and the results were placed into a contingency table shown below:

		1970			
		Agriculture	Developed	Forest	TOTAL
2000	Agriculture	29	2	7	38
	Developed	2	3	8	13
	Forest	5	2	62	69
	TOTAL	36	7	77	120

One can see the effect of urbanization, as the number of points in developed land has almost doubled between 1970 and 2000.

5. Perform a chi-square analysis of the land use change over the three decades and determine if changes in land cover follow a random distribution, or if land use changes follow a non-random pattern. For this exercise you will have to determine the expected number of points for each land use class, and then determine an overall chi-square statistic and p-value.

Contingency Tables - Soil Classifications

A random sample of 40 points on a soil series map was categorized by the soil order (Entisols, Alfisols, and Spodosols) and a modified drainage classification (well drained, moderately drained, and poorly drained). Certain soil types are preferred for different uses. For example, septic systems are most efficient on a well drained soil, while moderately drained soils work well for some agricultural purposes.

6. Perform a chi-square analysis on the contingency table below, and interpret the results. For this exercise you will have to determine the expected number of points for each soil order and drainage classification, and then determine an overall chi-square statistic and p-value.

	Well drained	Moderately drained	Poorly drained	Totals
Entisols	3	0	3	6
Alfisols	5	1	5	11
Spodosols	9	8	6	23
Totals	20	10	16	40

Workbook for Statistical Problem Solving in Geography

Chapter 14
Point Pattern Analysis

Major Goals and Objectives

In chapter 14 of *An Introduction to Statistical Problem Solving in Geography* you learned how to determine if spatial autocorrelation exists for point data by computing the nearest neighbor statistic and the variance to mean ratio (quadrat analysis). The data sets in the following exercises are intentionally small to allow you to compute both the nearest neighbor statistic and the variance-to-mean statistic by hand.

Problems and Exercises

Part I – Nearest Neighbor Analysis Colleges in Connecticut

In the United States, community colleges are primarily two-year public institutions providing higher education, career training, and certificates at costs much lower than four year colleges. Typically, community colleges are located by counties, but more rural areas may often have two or more counties share a community college.

The following maps show locations of the 12 community colleges in Connecticut, and the 4 comprehensive colleges along with the nearest neighbor distance in miles from one location to another (although a Public institution, the University of Connecticut is not a part of the Connecticut State University System). *Note that the State of Connecticut is 5,543 square miles.*

1. For this exercise, compute the nearest neighbor index for both the community colleges and the 4 year institutions and compare the results. Refer to table 14.2 in your textbook see a worked example for nearest neighbor calculation.

Community Colleges:

    State the null and alternative hypothesis:

    Compute the mean nearest neighbor distance:

    Calculate the standardized nearest neighbor distance:

    Calculate the test statistic and interpret the results:

Four Year Colleges:

    State the null and alternative hypothesis:

    Compute the mean nearest neighbor distance:

    Calculate the standardized nearest neighbor distance:

    Calculate the test statistic and interpret the results:

Workbook for Statistical Problem Solving in Geography   81

2. Compare the nearest neighbor statistic for the community colleges with the four year institutions (refer to Figure 14.5 in your textbook for an example). What observations can you make regarding the way the State located the public institutions for higher education?

Tornadoes in Kansas

Tornadoes with an F-Scale rating of 4 are considered a devastating, where well constructed houses are leveled, cars become airborne, and houses with weak foundations are blown away.

3. The following map shows the location of tornado touchdowns in Kansas between 1960 and 2013. Researchers are interested in knowing if the location of large tornado touchdowns (F-Scale 4 and above) are random, clustered, or dispersed. Note: the State of Kansas is 82,276 sq. mi.).

   State the null and alternative hypothesis:

   Compute the mean nearest neighbor distance:

   Calculate the standardized nearest neighbor distance:

   Calculate the test statistic and interpret the results:

hyyu

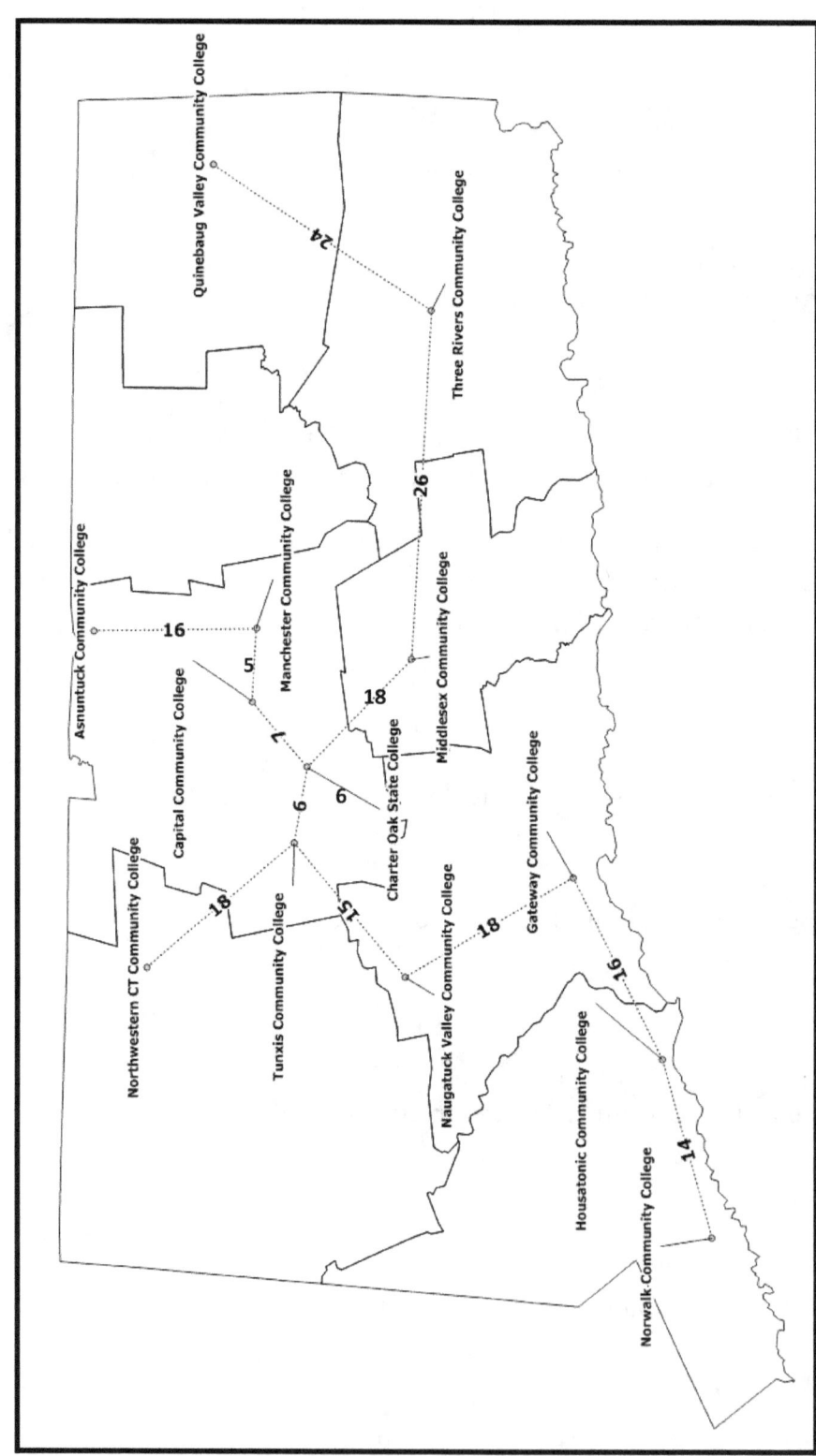

Community Colleges in the Connecticut University System

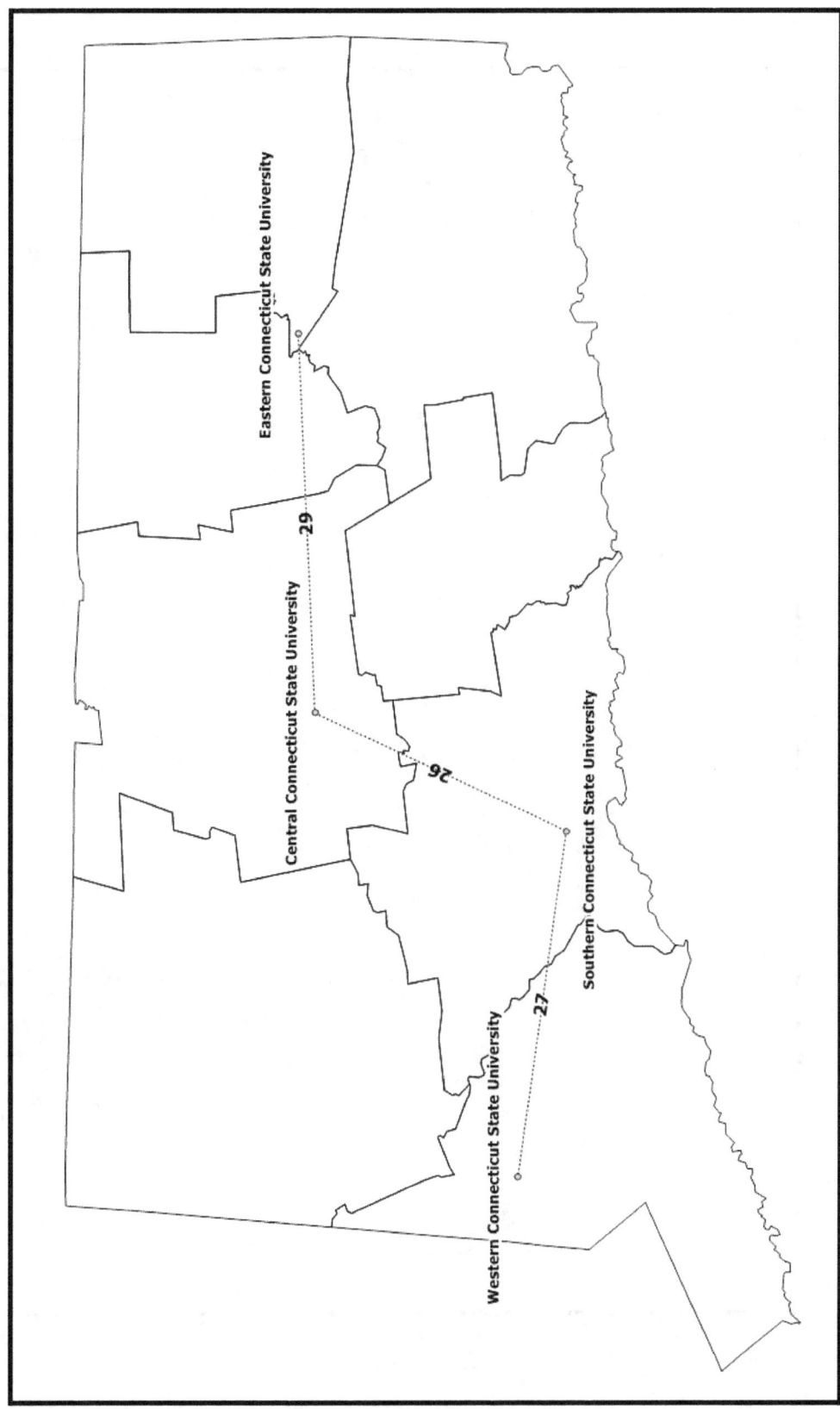

Four Year Colleges in Connecticut State University System

Workbook for Statistical Problem Solving in Geography

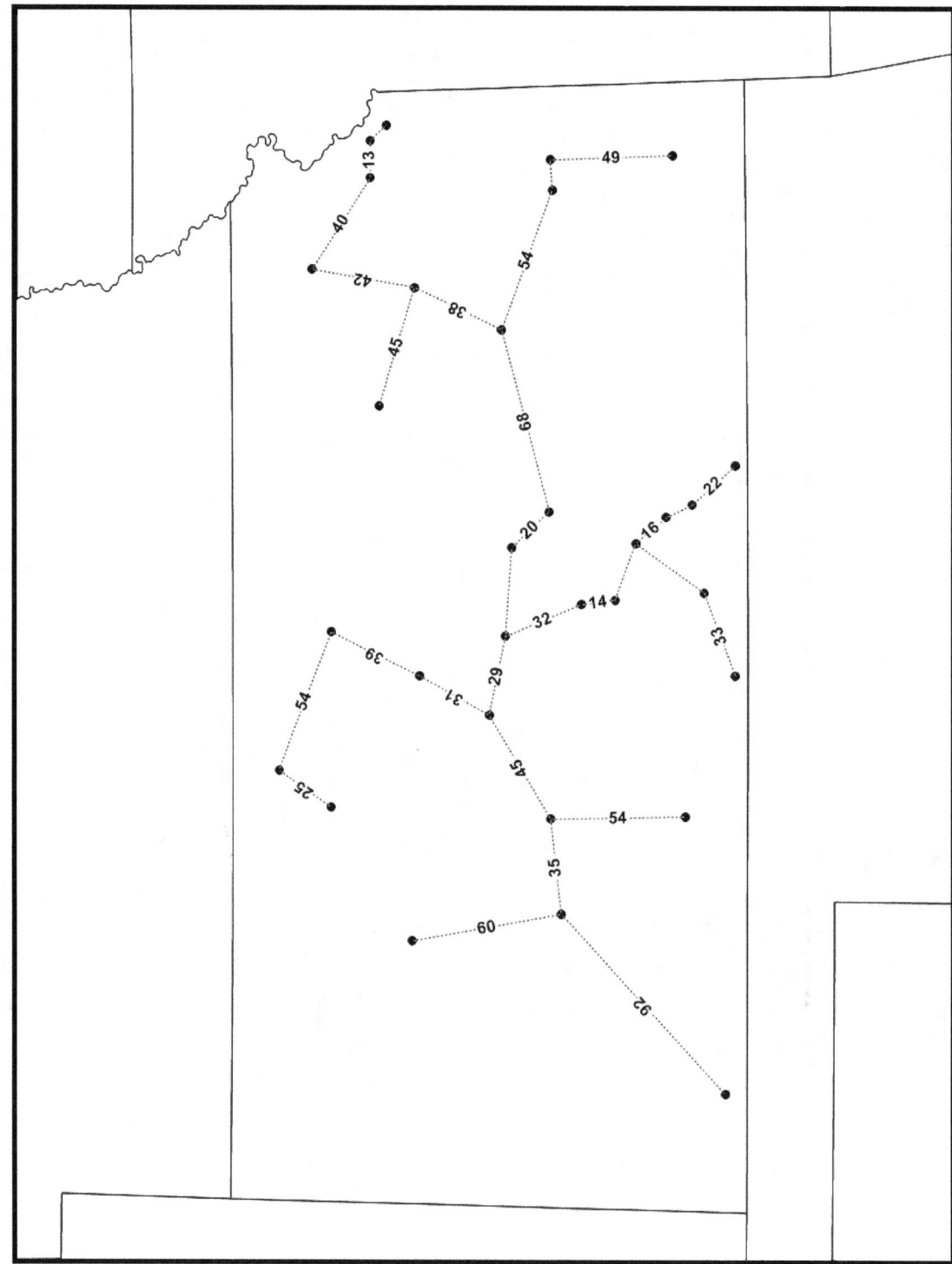

Tornadoes in Kansas 1965 - 2013 with an F-Scale Greater than 3

Workbook for Statistical Problem Solving in Geography

Part II – Quadrat Analysis - Hurricane Sandy Damage Assessment

Hurricane Sandy struck the east coast of the United States in 2012, and was the second costliest hurricane in United States history. Many communities were severely damaged, and years later, some the communities have yet to fully recover. Within days of the storm, FEMA began a rapid response to evaluate building damage using aerial imagery. Although a more comprehensive evaluate of damage would be carried out, these initial damage assessments were primarily used to gather immediate information about the extent of the damage.

This exercise will evaluate two beachside communities: Monmouth Beach, New Jersey and Bayville, New York. Both of these communities experienced damage, and a quick glance at the maps would indicate that the damage is not random.

4. The points on the map indicate locations of buildings that appeared to show inundation. Using Table 14.4 of your textbook as a guide, compute the variance-to-mean ratio for the damage locations. Please note that the scale of the map may make counting the individual points difficult. It is likely that you will miss one or two points here or there. However, because of the large number of points, missing a couple of locations will not have a very large impact on the overall results.

Monmouth Beach, New Jersey

State the null and alternate hypothesis:

Calculate the mean cell frequency:

Calculate variance of cell frequencies:

Calculate variance-to-mean ratio:

Calculate the test statistic and interpret the results:

Bayville, New York

State the null and alternate hypothesis:

Calculate the mean cell frequency:

Calculate variance of cell frequencies:

Calculate variance-to-mean ratio:

Calculate the test statistic and interpret the results:

5. Provide a comparison of the spatial distribution of damage to both communities.

Workbook for Statistical Problem Solving in Geography

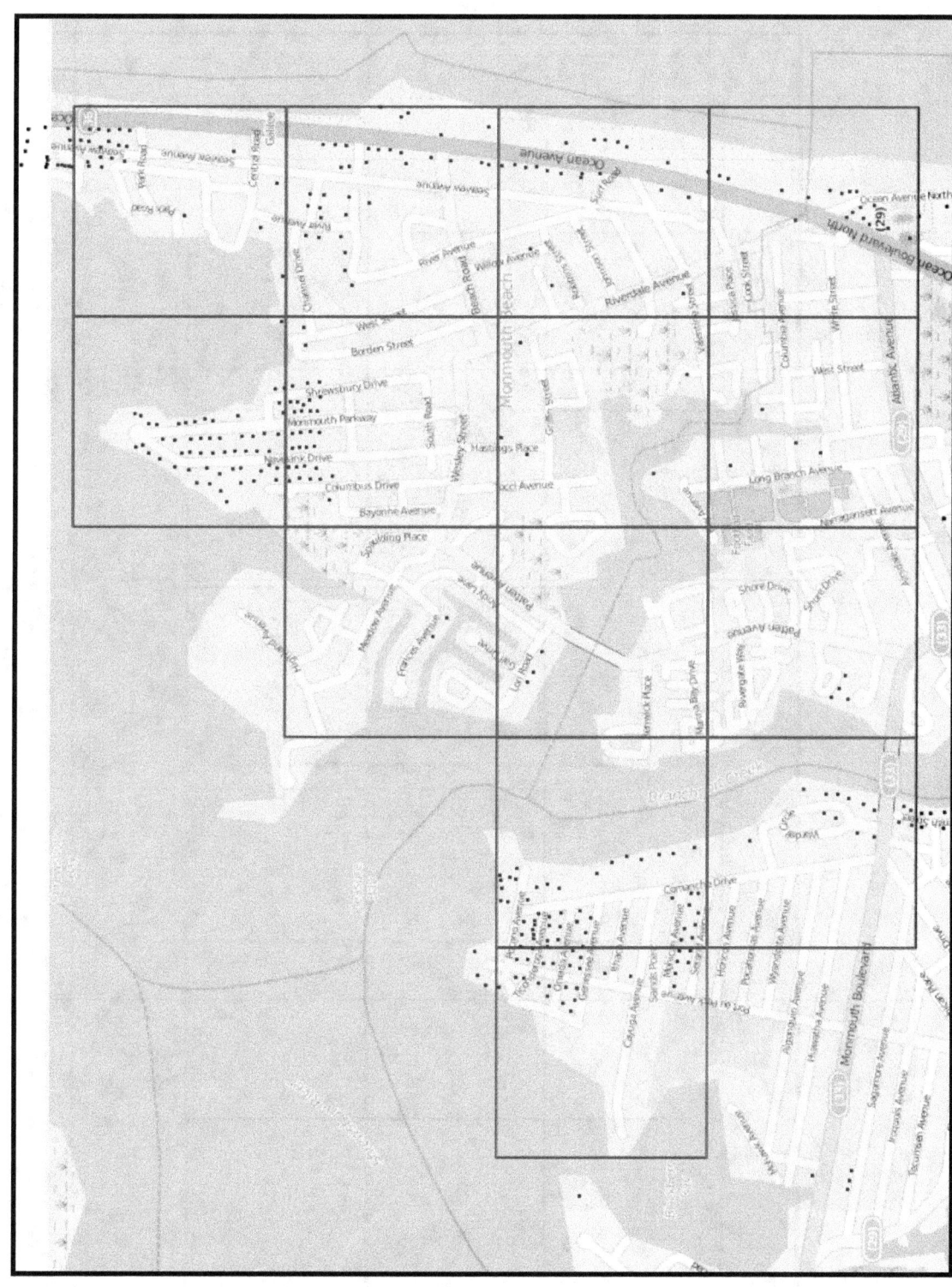

Initial Hurricane Sandy Damage - Monmouth Beach, New Jersey

Workbook for Statistical Problem Solving in Geography

Initial Hurricane Sandy Damage - Bayville, New York

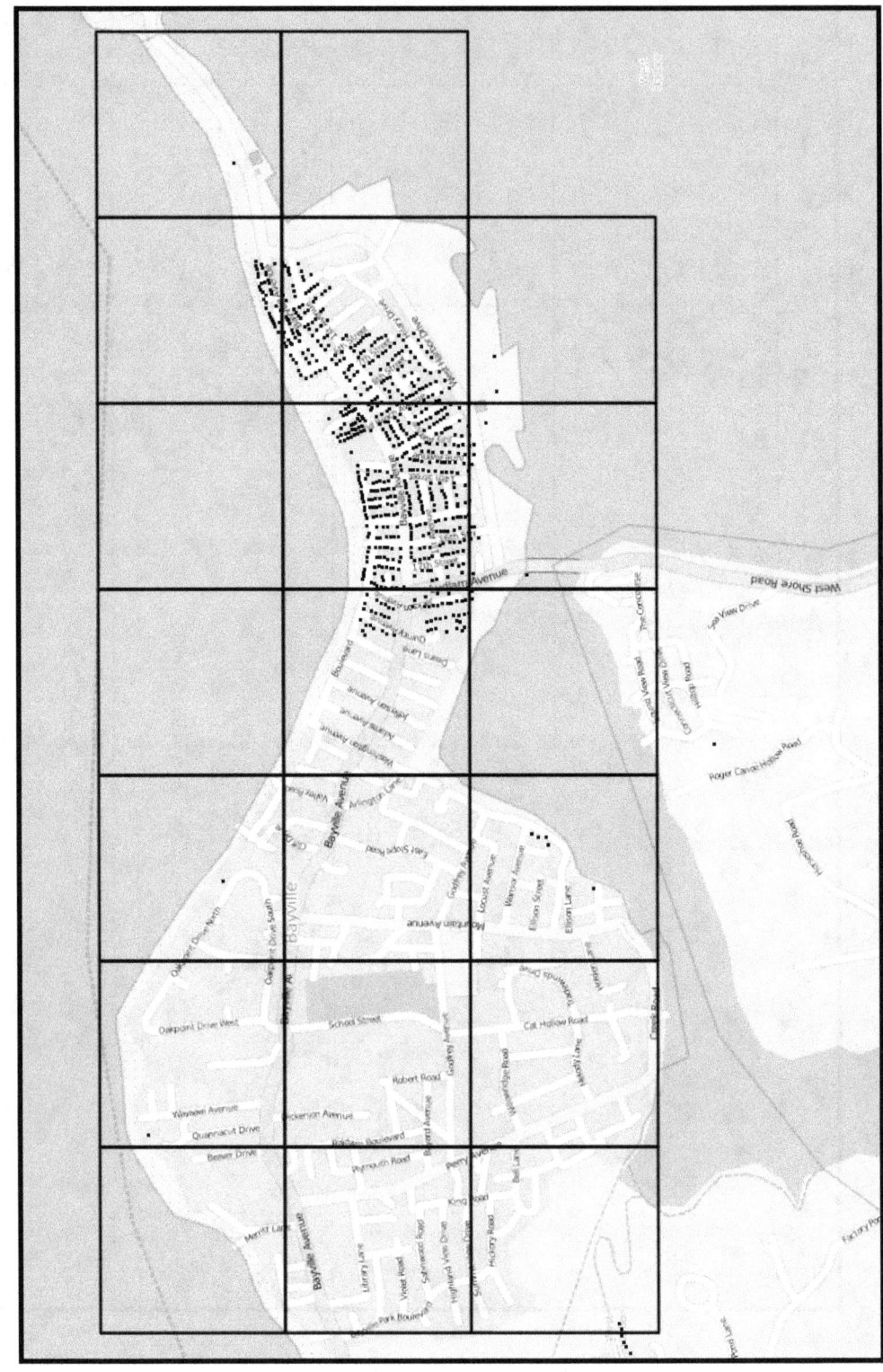

Chapter 15
Area Pattern Analysis

Major Goals and Objectives

In chapter 15 of *An Introduction to Statistical Problem Solving in Geography* you learned how to determine if spatial autocorrelation exists for area data by computing the *join count* analysis and *Moran's I* coefficient. In the following example, 2008 obesity values are shown for counties in the State of Connecticut. The data are represented as the actual percentage obese in each county as well as being shaded to indicate if the obesity values are above or below the state average.

The data set is intentionally small to allow you to compute both the join count and Moran's coefficient by hand. Refer to Table 15.4 in your textbook for the join count, and Table 15.6 for the Moran's coefficient.

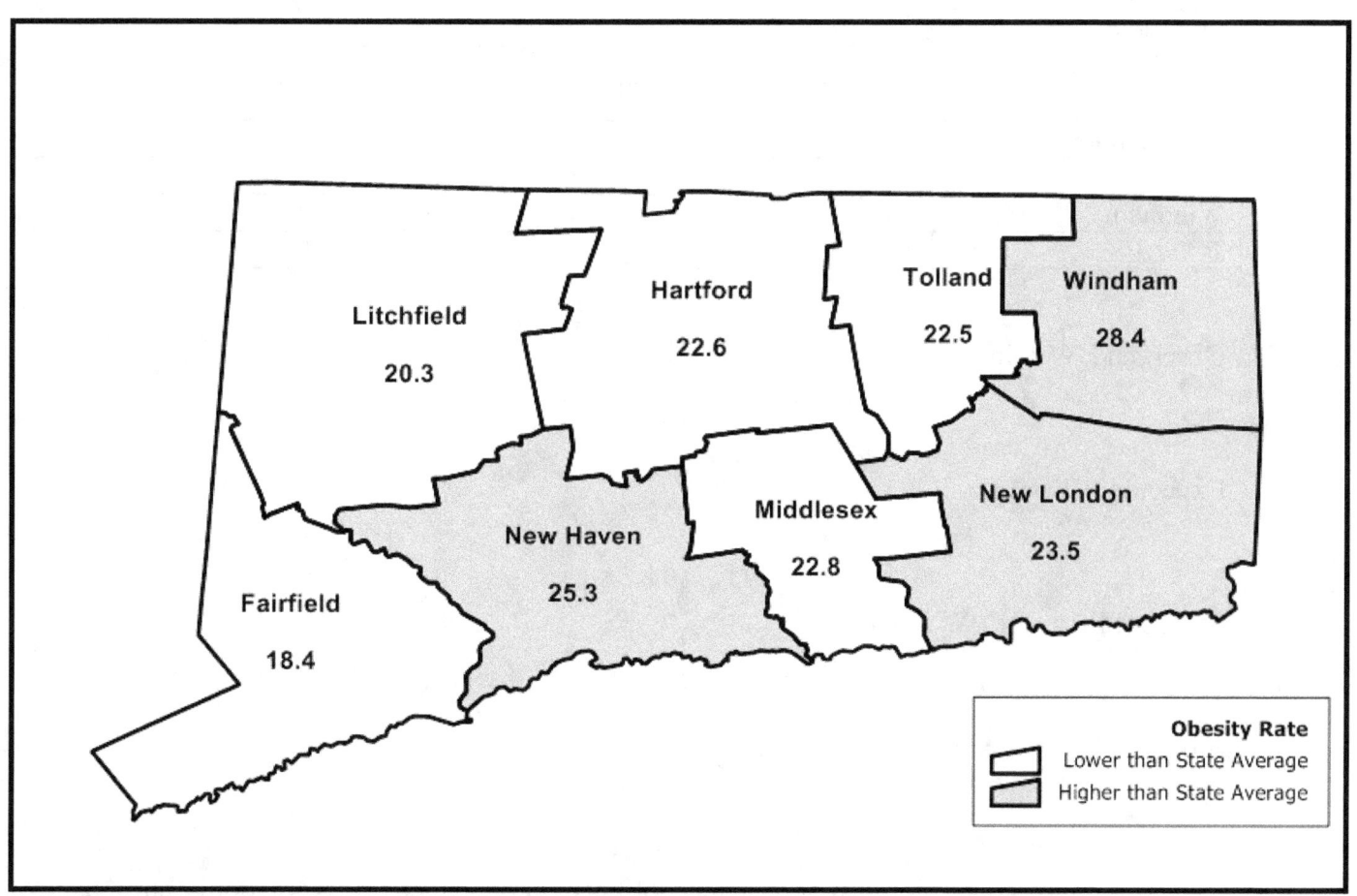

Obesity Rates in Connecticut (2008)

Join Count Analysis - Obesity Rates in Connecticut

1. Determine if spatial autocorrelation exists among the Connecticut counties that are above the State obesity rate - in this case, compute the BW join structure for the State.

a. State the null and alternate hypothesis:

b. Calculate the join count statistic as shown in Table 15.3 and Table 15.4 of your textbook.

County Name	Number of Links (L)	(L-1)	L(L-1)
Fairfield			
Hartford			
Litchfield			
Middlesex			
New Haven			
New London			
Tolland			
Windham			
TOTALS			

Observed BW Joins:

Expected BW Joins:

Standard Error for BW Joins:

Z Score:

2. Offer your conclusion as to whether obesity in Connecticut is random, clustered, or dispersed.

Morans I - Obesity Rates in Connecticut

3. Determine if spatial autocorrelation of obesity exists among the Connecticut counties. Follow the example from Table 15.6 in your textbook to assist you.

a. State the null and alternate hypothesis:

b. Calculate the join count statistic as shown in Table 15.6 of your textbook.

Join Number	x_i	$(x_i - \bar{x})$	x_j	$(x_j - \bar{x})$	$(x_j - \bar{x})(x_i - \bar{x})$
1					
2					
3					
4					
5					
6					
7					
8					
9					
10					
11					
12					
13					
TOTALS					

Expected I:

I:

Var$_i$:

Z:

Chapter 16
Correlation Analysis

Major Goals and Objectives

In this chapter you learned how to distinguish various directional relationships between variables and also evaluate the strength of those relationships. You also learned how to conduct a correlation analysis for both parametric (Pearsons) and ranked (Spearmans) data, and interpret the results.

In this exercise you will evaluate the strength of the relationship between datasets as it relates to some controversial subjects. The choice of these controversial subjects is not an endorsement of any particular viewpoint, but simply to provide you with a fun opportunity to quantitatively assess data for which many citizens have already formed an opinion on without having actually done any quantitative analysis. As you will recall from the textbook, *correlation does not necessarily imply causation*, and there are often other variables that might provide further insight into the situation. Therefore, consider these datasets a starting point in the discussion of controversial ideas.

<u>Pearsons Correlation Coefficient - Poverty and Births to Teenage Mothers</u>

The argument goes something like this: higher rates of teen pregnancy lead to poverty within a community. Another argument says: high rates of poverty within a community lead to teen pregnancy. Perhaps a better question to answer before engaging in the debate is whether there is even a relationship between teen pregnancy and poverty.

1. While correlation analysis does not distinguish about the form of the relationship (i.e. what factor causes another), it does allow us to determine if there is any association among the variables to begin with. For this exercise we will revisit tab *Chapter 2 - US Data*, and compare the **Births to Teenage Mothers (2009)** to **Poverty Rate (2010)** to determine if a statistically significant relationship exists at the statewide level for these two variables.

a. State the null and alternate hypothesis:

b. Calculate the Pearsons Correlation Coefficient:

c. Offer your conclusion as to whether there is a relationship between births to teenage mothers and poverty rate.

Pearson Correlation Coefficient - Guns and Murders

2. The argument goes something like this: the availability of guns within a community increases the likelihood of murders within that community. Another argument says: the presence of legal guns in a community deters murders. The issue of gun violence is very emotional, and most recognize there are no easy answers to the problem. However, while most everyone has an opinion on the issue, very few have considered a quantitative assessment as to whether there is any relationship among gun ownership and murders. For this exercise consider tab Chapter 16 - Guns Murders, and compare the **gun murders rate per 100,000 people** to the **percent gun ownership** at the statewide level.

a. State the null and alternate hypothesis:

b. Calculate the Pearsons Correlation Coefficient:

c. Offer your conclusion as to whether there is a relationship between gun ownership and the murder rate by gun violence.

Pearson Correlation Coefficient - Income Taxes and Unemployment

3. The argument goes something like this: *higher taxed states stifle employment opportunities and increase the unemployment rate*, while another argument would indicate that higher taxed states provide more resources that might help lower unemployment. Obviously, many factors contribute to unemployment in a state. Nonetheless, before engaging in the argument, it might be beneficial to determine if there is a relationship among taxes and unemployment at the statewide level. For this exercise, consider tab *Chapter 16 Unemployment - Taxes* to determine if a relationship between these variables exists in a statistically significant way.

a. State the null and alternate hypothesis:

b. Calculate the Pearsons Correlation Coefficient:

c. Offer your conclusion as to whether there is a relationship between income tax rates and unemployment.

Spearman Rank Correlation - Corporate Taxes and Unemployment

4. Another argument one might make is that irrespective of State income tax, corporate taxes have an impact on unemployment rates - the higher the tax, the higher the unemployment. Similar to the other examples, there are many factors that affect unemployment besides taxes. Also compounding the difficulty in evaluating corporate taxes and unemployment is that fact that many states have different rates based on the revenue for the corporations. Nonetheless, a good starting point in the debate is to ascertain if there is any relationship between corporate tax rates and unemployment. For this exercise, consider tab *Chapter 16 Unemployment - Taxes* to determine if a relationship between these variables exists in a statistically significant way. Due to the multiple tax rates for a state, only the highest rate for the state is shown. Therefore, our confidence in the data being an accurate reflection of the corporate tax rate is someone hindered, necessitating a non parametric analysis of the data.

To complete this analysis, you will need to rank each State in terms of its corporate tax rate and unemployment.

a. State the null and alternate hypothesis:

b. Calculate the Pearsons Correlation Coefficient:

c. Offer your conclusion as to whether there is a relationship between income tax rates and unemployment.

Chapters 17 and 18
Regression Analysis

Major Goals and Objectives

In this chapter you learned what regression is, and how to conduct a regression analysis, taking things like multi-colinearity into account. As part of the regression analysis, you learned how to interpret the regression equation, explain the variation, r^2 value, and p-value. Finally, you learned how to use the regression coefficients to perform a *what-if* scenario as part of a predictive model.

While Table 17.1 illustrates how to compute the slope and aspect for simple linear regression, calculating these values by hand is somewhat lengthy. Therefore, it is suggested that you use Microsoft Excel, Minitab, or another statistical program to compute the regression results, and focus your time on interpreting the tables.

Further, multiple regression is a multivariate least-square technique that is beyond a reasonable ability to compute by hand. Therefore, to complete these examples you will need software to compute the regression analysis. While Microsoft Excel is easy to use, it lacks some of the more important features such as computation of the sequential sum of the squares to determine the *leverage* that a variable might be contributing to the solution. In addition, due to the multi-collinearity, a stepwise regression will help determine the best variables to use in the solution, something Microsoft Excel cannot perform. Therefore, if possible, utilize a software package like Minitab, SAS, or R that is better suited for regression analysis.

Problems and Exercises

Simple Linear Regression - Water Temperature and Dissolved Oxygen

In chapter 10 of your workbook we explored the relationship of dissolved oxygen (DO) and temperature. The general theory is that higher water temperatures yield lower DO concentrations. However, in chapter 10 we only explored whether there were any differences between warm and cold months and wet and dry months. It is possible that fluctuations in water temperature throughout the day might present stress in some fish species.

1. In this example, we are interested in the nature of the relationship between water temperatures and DO concentration. Tab *Chapter 18 Temp and DO includes* 44 samples from a USGS monitoring station from the White River in Auburn, Washington. For this exercise, perform a simple linear regression to determine the nature of the relationship between water temperature and dissolved oxygen content in the White River.

For the results of your analysis, interpret the following items:

a. the equation of the least-squares regression line, including the *y* intercept and slope *b*.

b. the explained and unexplained variation.

c. the coefficient of determination.

d. the standard error of the estimate

e. the test statistic for r^2 and the p-value.

f. indicate what these results mean in terms of the relationship between dissolved oxygen and water temperature. Also, consider how the predictive nature of the regression equation can determine what kinds of temperatures might present oxygen depletion for some species of fish (under 5 mg/L). Recognize, however, that your prediction will fall outside the bounds of the input data.

Simple Linear Regression - Biomass and Tree Diameter

According to the Food and Agricultural Organization (FAO) of the United Nations:

Biomass estimates for forests of tropical countries, in particular, are needed because globally they are undergoing the greatest rates of change and reliable biomass estimates are few. Their biomass and C content is generally high, which influences their role in the global C cycle. Further, tropical forests have the greatest potential for mitigation of CO_2 through conservation and management.

Additionally, the FAO states that estimating the biomass of tropical forests is very relevant for issues related to global change, including the role of tropical forests in global biogeochemical cycles, especially the carbon cycle and its relation to the greenhouse effect.

One approach is involves estimating the biomass per average tree of each diameter (diameter at breast height, dbh) class within a stand. Those estimates may then be multiplied by the number of trees in the class to estimate total biomass within the forest.

2. Tab *Chapter 18 – Biomass* includes dbh and biomass values for a specific tree species within two dry zone forests. Perform linear regression to explore the relationship between biomass and tree density.

For the results of your analysis, interpret the following items:

a. the equation of the least-squares regression line, including the y intercept and slope b.

b. the explained and unexplained variation.

c. the coefficient of determination.

d. the standard error of the estimate

e. the test statistic for r^2 and the p-value.

f. consider the predictive nature of the regression equation to determine how biomass for an entire forest might be predicted from a few samples.

Simple Linear Regression - Drug Arrests and Percent Owner Occupied Homes

Urban geographers believe that communities with a large percentage of owner occupied homes have less drug problems, as homeowners tend to be less tolerant of drug activity within their neighborhood. Tab *Chapter 17 - Drugs and Housing* lists the number of drug arrests and the percent owner occupied homes by census block group in Salisbury, MD.

3. Perform a regression analysis to determine the relationship between drug arrests and percent owner occupied homes for this community.

For the results of your analysis, interpret the following items:

a. the equation of the least-squares regression line, including the *y* intercept and slope *b*.

b. the explained and unexplained variation.

c. the coefficient of determination.

d. the standard error of the estimate

e. the test statistic for r^2 and the p-value.

f. imagine you were asked by the City council to provide testimony as it relates to home ownership and drug arrests. What might you tell the council? For this answer, think about the predictive power of the regression table and how increasing home ownership impacts the drug crime within a community.

Multiple Regression - Housing Values

Urban and economic geographers often want to determine a mathematical relationship to predict home sales. This is a very difficult relationship to understand, as the value of a home is more than the simple statistical data that describe the home. Personal preferences oftentimes determine what someone will pay for a home. However, some factors such as the age and size of the home, the quality of the surrounding neighborhood and how far it is from the water may place a predictive role in determining the sale price of a home.

Tab *Chapter 18 - Housing Prices*, includes a sample of 40 home sales in the City of Annapolis, MD in 2005. In addition to the sale price, characteristics such as the age of the home, the size of the property, and the square footage of the home were collected. One might hypothesize that newer and larger homes would have a higher sales price. In addition, a GIS analysis was conducted to compute the distance of the property to the water and the average assessed value of neighboring properties within 500', as one might hypothesize that homes nearer to the water and part of an already expensive neighborhood would demand a higher sales price.

4. Perform a multivariate regression analysis on the house sale data and interpret the following:

a. the equation of the least-squares regression line, including the significance of each coefficient.

b. the standard error of the estimate

c. the test statistic for r^2 and the p-value for each coefficient.

d. imagine you worked for an Internet-based startup company looking to serve the housing market by providing potential homebuyers with a formula for determining what they should expect to pay for a house, given particular parameters. Use the regression equation to illustrate what variables should be considered in the Annapolis example, and provide a hypothetical scenario for a range of values that a user might offer on a house given some of the significant parameters.

Multiple Regression - Crop Yield

An experimental farm in upstate New York collected 24 soil samples and corresponding yield values, and shown in tab *Chapter 18 - Crop Yield*. The soil parameters include organic matter, Potassium (K), pH, and Phosphorous (P).

5. Perform a multivariate regression analysis on the crop yield data to determine the relationship that the soil parameters have on crop yield, and interpret the following:

a. the equation of the least-squares regression line, including the significance of each coefficient.

b. the standard error of the estimate

c. the test statistic for r^2 and the p-value for each coefficient.

d. imagine you are an agricultural consultant, helping local farmers determine the most suitable soil conditions for farming. What observations would you make regarding crop yield based on the soil parameters?

www.ingramcontent.com/pod-product-compliance
Lightning Source LLC
Chambersburg PA
CBHW081730170526
45167CB00009B/3765